CONTENTS

T308

En
m

prepared for the course team
by Stephen Burnley, David Cooke and Toni Gladding

03/07

This publication forms part of an Open University course T308 *Environmental monitoring, modelling and control*. Details of this and other Open University courses can be obtained from the Student Registration and Enquiry Service, The Open University, PO Box 197, Milton Keynes, MK7 6BJ, United Kingdom: tel. + 44 (0)870 333 4340, e-mail general-enquiries@open.ac.uk

Alternatively, you may visit the Open University website at http://www.open.ac.uk where you can learn more about the wide range of courses and packs offered at all levels by The Open University.

To purchase a selection of Open University course materials visit http://www.ouw.co.uk, or contact Open University Worldwide, Michael Young Building, Walton Hall, Milton Keynes MK7 6AA, United Kingdom for a brochure: tel. + 44 (0)1908 858793; fax + 44 (0)1908 858787; e-mail ouw-customer-services@open.ac.uk

This course has been printed on Savannah Natural Art™. At least 60% of the fibre used in the production of this paper is bagasse (the fibrous residue of sugar cane, left after the sugar has been extracted) and the balance is softwood fibre which has undergone an oxygen bleaching process.

The Open University,
Walton Hall, Milton Keynes
MK7 6AA

First published 2005. Second edition 2007.

Edited and designed by The Open University.

Typeset in India by Alden Prepress Services, Chennai.

Printed and bound in the United Kingdom by Hobbs The Printers Limited, Brunel Road, Totton, Hampshire SO40 3WX.

ISBN 978 0 7492 1864 5

2.1

LEARNING OUTCOMES

The Wastes block of T308 builds on the material of the Open University second level course T210 *Environmental control and public health,* taking you to greater depth in specific areas of wastes management.

After studying the text and working through the associated questions, exercises and the assignment you should be able to:

- analyse critically the economic, technical and environmental issues involved in modern waste management
- apply the principles of integrated solid waste management to devise and assess the impacts of integrated waste management strategies
- develop and use computer and other models to assess waste management operations and systems
- prepare material for an environmental impact assessment relating to wastes management issues
- appreciate, specify and analyse the elements of an environmental impact analysis in wastes management
- identify those areas of the environment upon which waste management impinges and outline the steps taken to minimise adverse effects.

This block provides opportunities for you to develop and demonstrate the following learning outcomes.

Knowledge and understanding:

You should have a knowledge and understanding of:

1 United Kingdom waste management policy, legislation and regulation (SAQs 1, 2, 11 and 30, Exercise 1)

2 the quantities and characteristics of the different wastes produced in the UK (SAQs 2–4, 10 and 17)

3 the environmental impacts relating to the collection of wastes and the operation of transfer stations, materials reclamation facilities, composting plants, mechanical biological pre-treatment (MBP) plants, incinerators and landfills (SAQs 3, 5, 8, 11 and 14–29)

4 the operator and public health impacts of the collection and treatment of solid wastes (SAQs 6, 7, 9, 12 and 13)

5 the extent to which different waste management options can contribute to achieving local, national and European waste management requirements (Computer Activities 5 and 6)

6 the principles of integrated solid waste management (Computer Activities 5 and 6)

7 techniques for comparing the environmental impacts of different waste management options (SAQ 5, Computer Activity 6).

Thinking (cognitive) skills

You should have the skills to:

1 use the principles of integrated solid waste management to devise waste management strategies at the district and county levels (Computer Activities 5 and 6)

2 discuss the health and environmental implications of the main waste collection, processing and disposal activities (SAQs 5–9, 13)

3   perform and interpret landfill water balance calculations (SAQs 21–23, Computer Activity 2)

4   model the performance of landfill site lining systems (SAQs 24 and 25, Computer Activity 2)

5   model the performance of incinerator flue gas scrubbing systems (SAQs 15 and 16)

6   perform detailed assessments of waste management options and systems (Computer Activities 5 and 6).

Professional skills

You should have the skills to:

1   produce written articles for expert and non-expert audiences to analyse waste management problems and recommend solutions (Tutor-marked assignments, Project)

2   assess the environmental impacts relating to wastes management of a major infrastructure project and write the waste-related sections of an environmental impact assessment statement (Tutor-marked assignment, Project).

Key skills

You should have the skills to:

1   critically review conflicting articles in the academic and professional press relating to the benefits and adverse impacts of different waste management options (Exercise 3)

2   develop and use computer models to assess the financial and environmental impacts of individual waste management options and integrated combinations of options (Computer Activities 5 and 6, Project).

1   INTRODUCTION

1.1   Aims of this block

Imagine you are a waste management specialist in a local authority with the responsibility for the strategic forward planning of municipal solid waste management in your area.

What choices are open to you as you plan the future of waste management? How can you achieve the sometimes conflicting aims of best practicable environmental option (BPEO) and apply the waste management hierarchy while still meeting your targets for recycling and diversion from landfill? What opportunities exist for reduction and reclamation of waste in both the public and private sectors, and how economic are they? What problems will you have in ensuring that waste management in your area is carried out in an environmentally acceptable manner? How will you convince a sceptical public that their health won't be harmed by your landfill or incinerator?

Finding satisfactory answers to these and the many other questions you need to ask requires the involvement of many disciplines and the balancing of many opposing factors. As you will see as you work through the Wastes block, advances have been made in the way wastes are managed and standards are continually increasing. However, there are still areas of uncertainty in interpretation of available scientific evidence, and other areas where environmentally desirable practices are not necessarily implemented, principally because of economic factors.

It is the intention of this block to give you the necessary background to help you to be in a better position to decide whether waste management in the UK is currently carried out as well as it might be, and what options are open for its improvement. You will also study various techniques for modelling waste management activities and their impact on the environment. This forms an essential part of the environmental impact statement that you will write for the course project.

The block can be divided into three parts:

1   Sections 1 and 2 cover the background information that you need to begin this block.

2   Sections 3 and 4 build on this background knowledge of the technology of the main waste management options by considering the environmental impacts of these options. This involves the techniques used to measure the fugitive emissions from waste collection and treatment processes (the measurement of emissions from chimneys, vents and flares is covered in Block 2: *Managing air quality*). The wider impacts of waste composting and recycling are then considered in terms of their environmental benefits and risks.

One of the main environmental impacts of incineration arises from the combustion gas emissions. Section 5 considers the composition of these gases, the discharge standards and the operation of flue gas treatment systems. Incineration can be a highly emotive area; virtually all proposals for incinerators are opposed on environmental and health grounds by local, national and multinational organisations. Section 5 concludes by making a critical review of some of the conflicting literature on incineration.

Section 6 considers landfill. Although the European Commission's directive on the landfill of waste (the Landfill Directive) is resulting in a reduction in the landfilling of some types of waste, landfill will be required for the foreseeable future and its potential environmental impacts will last for many decades after the closure of a landfill site. The formation of leachate and landfill gas are reviewed, and you will develop and use computer models to study their formation and environmental impact.

Finally, Section 7 reviews a number of novel waste management options that may be adopted in the future.

3   The third part begins by considering ways of evaluating the environmental impacts of different activities or processes using the technique of life-cycle assessment (LCA). The discussion is then broadened to include two tools to assist in decision making in general: multi-criteria analysis (MCA) and multi-criteria decision analysis (MCDA).

The final part of Section 8 concentrates on modelling integrated solid waste management (ISWM) systems. ISWM strategies are generally based on combinations of recycling, composting, incineration and landfill. Models are used to predict the effectiveness of a given strategy in terms of compliance with the Landfill Directive and national recycling and recovery targets. The costs of implementing the strategy are also considered by the model.

Throughout this block I have concentrated on household and municipal waste. At first sight, it may seem strange to focus so much attention on what amounts to about 15% of the UK's controlled waste stream. However, as I pointed out in T210 Block 4: *Wastes management* this is principally due to policy and legislative drivers. The main impact of the Landfill Directive is related to the need to divert biodegradable municipal waste from landfill. Also the national recycling targets relate to municipal and household wastes. In addition, municipal waste has many distinctive problems that do not apply to other wastes:

- It is very diffuse, being generated in some 25 million households over an area of 244 000 km^2.

- It contains a much wider range of materials than any given industrial waste; and is less regulated in that householders can pretty much throw out what they want, even hazardous items.

- The producers (i.e. the householders) do not have direct legal responsibility for its disposal.

- The producers do not pay for waste collection and disposal directly.

Nevertheless, by studying municipal and household wastes, you will come across all the main problems faced when managing commercial, industrial and agricultural wastes. You will also develop an understanding of the technologies that can be used to manage these other wastes.

1.2   Background knowledge

This block assumes that you already have a basic knowledge of wastes and waste management policy and practice in the UK. Many of you will have achieved this through study of The Open University's second level course T210 *Environmental control and public health* or its predecessor T237. Some of you will have gained this knowledge through studying other courses or through your work. In any event, you should have access to Block 2: *Wastes management* of T210.

The following paragraphs and SAQs summarise the material that you should be familiar with. Waste management is a rapidly changing field so you should also read the relevant sections of the Legislation Supplement and check the T308 website to keep abreast of recent developments.

1.2.1   Waste policy and regulation

In common with most environmental pollution control policy, the UK's waste policies are being developed and implemented in parallel with EU-wide policies. This is being done with the intention of making waste management more 'sustainable'. In this context, I have defined sustainable to mean:

> minimising the consumption of non-renewable natural resources, minimising environmental pollution and ensuring that any pollution caused is dealt with in a way that leaves the minimum possible legacy for future generations to deal with.

More specifically, those charged with managing wastes must apply the principle of best practicable environmental option (BPEO) which I will discuss further in Section 8.1. This means ensuring that the waste management option selected for a particular waste provides the least environmental damage (or ideally, the greatest environmental benefit). The EU waste management hierarchy provides a guide to help define BPEO. The hierarchy, in order of desirability, is:

- reduce the amount of waste to be managed
- reuse waste products and materials
- recover value from wastes by recycling, composting and energy recovery
- dispose of wastes by landfill when there is no other option.

To achieve these aims, the governments and assemblies of the UK have produced waste management strategies which set targets for waste recycling and recovery to be achieved by the local authorities. Those for England are summarised below.

Household waste recycling and composting:

> 25% by 2005
>
> 30% by 2010
>
> 33% by 2015

Municipal waste recovery (recycling, composting and incineration with energy recovery):

> 40% by 2005
>
> 45% by 2010
>
> 67% by 2015

If you are uncertain about the definitions of household and municipal wastes see Section 3 of T210 Block 4: Wastes management.

Additionally, as one of the Government's Local Authority Best Value Performance Indicators (BVPI), each English local authority has been set statutory recycling/composting rates. For the year 2005/06 these ranged from 18% to 30% depending on the rate achieved in 1998/99. For 2007/08 authorities with an 18% target in 2005/06 will have to reach 20% and those who have achieved higher rates will not see any increase in their target.

EXERCISE 1

Find the website of your national organisation responsible for waste strategy development (site details are given on the T308 website) and spend a few minutes finding out what targets it has set for waste recycling and recovery.

Furthermore, the Landfill Directive, which aims to minimise pollution due to landfilling, calls for the following reductions in the amount of biodegradable municipal solid waste (MSW) sent to landfill. These dates include a four-year delay allowed for countries with a high reliance on landfill.

> To 75% of the amount produced in 1995 by 2008
>
> To 50% of the amount produced in 1995 by 2013
>
> To 35% of the amount produced in 1995 by 2020.

This means that, by 2020, the UK will have to divert at least 15 million tonnes a year of MSW from landfill compared with the amount landfilled in 1995 (Burnley et al., 1999). There is much debate about how this can be achieved, but it will certainly mean a massive expansion in waste recycling, composting, incineration and other novel management methods.

Now use your background knowledge of waste or the material in T210 to answer the following SAQs.

SAQ 1

Despite campaigns to reduce the amount of wastes produced and the rising cost of waste management, quantities of MSW are increasing. Give a number of reasons why this may be the case.

SAQ 2

Civic amenity waste differs in composition from household-collected MSW. What differences would you expect to find and why is this so?

1.2.2   Waste quantities, composition and disposal

The Department for Environment, Food and Rural Affairs (DEFRA) estimates that the UK produces around 420 million tonnes of waste a year. This is broken down as shown in Figure 1 below.

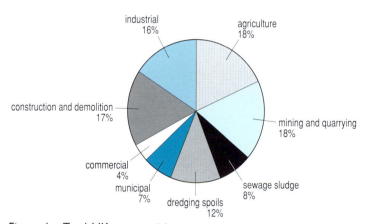

Figure 1   Total UK waste arisings

Mining and quarrying waste, sewage sludge and dredging spoils are not classed as 'controlled waste' owing to their relatively low potential for causing environmental pollution. This means that their disposal is not regulated under the waste licensing regime by the national regulator. Until May 2006 agricultural waste was not a controlled waste, but now all agricultural wastes (apart from crop residues and animal manures used as fertilisers) are regulated in the same way as any other industrial or commercial wastes.

MSW is defined as the solid wastes produced by households and similar waste that is collected by, or on behalf of, a local authority. The main components of MSW are shown in Figure 2. Most of the Landfill Directive and national recycling/recovery targets refer to MSW so this material has been the subject of many studies to determine its composition. Table 1 gives some typical values for the composition of the main components of household waste (dustbin waste plus recyclable materials collected from individual households) and civic amenity site waste. These values were taken from a survey of MSW carried out in Wales in 2003 (Poll, 2004). Collectively these two streams comprise about 72% of MSW.

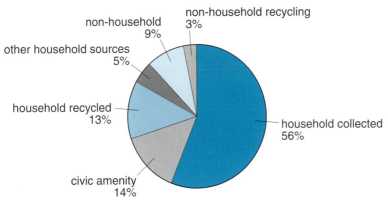

Figure 2   Municipal solid waste components (England 2001–02). Note: Other household sources include bulky waste collections, street sweepings and litter (Source: DEFRA, 2004a)

Table 1   Household-collected waste composition

Category	Household-collected waste (%)	Civic amenity site waste (%)
Paper and card	25	8
Plastics	11	3
Textiles	2	2
Other combustible material (nappies, wood, leather, etc.)	8	29
Non-combustible material (ceramics, rubble, etc.)	3	20
Glass	7	2
Kitchen and garden wastes	36	19
Metals	5	8
Electrical items	<1	7
Hazardous items	<1	1
Unclassified material less than 10 mm in size	3	1

Source: Poll (2004).

The chemical composition of household-collected waste from across the UK was determined as part of the National Household Waste Analysis Project (NHWAP) in 1993. The main findings are summarised in Table 2. You should note that the composition will almost certainly have changed since 1993; in particular, the heavy metal concentrations will be lower.

Table 2   Chemical composition of household collected waste

	Units	Mean	Minimum	Maximum
Moisture	%	34.18	27.5	41.2
Ash	%	21.64	13.61	33.02
Gross calorific value	MJ kg^{-1}	9.7	6.4	13.3
Fixed carbon	%	5.5	3.6	7.3
Volatile matter	%	38.7	28.9	48.9
C	%	24.3	19.2	28.9
H	%	3.4	2.7	4.4
N	%	0.6	0.33	1.2
S	%	0.1	0.05	0.55
Cl	%	0.3	0.13	0.97
Br	%	0.01	0.001	0.048
F	%	0.01	0.009	0.04
Al	%	1.8	0.7	11
Hg	ppm	0.10	0.01	1.2
Si	%	2.83	0.53	6.59
Na	ppm	5 977	1 367	16 399
Mg	ppm	1 868	811	3 359
K	ppm	2 936	1 214	5 423
Ca	ppm	13 208	6 422	47 041
Cr	ppm	254	24	583
Mn	ppm	582	232	4 694
Fe	ppm	45 039	3 217	82 230
Ni	ppm	39	17	120
Cu	ppm	1 895	27	13 584
Zn	ppm	1 442	149	10 117
As	ppm	14	1.8	204
Mo	ppm	3.3	1.8	9
Ag	ppm	1.7	0.2	16
Cd	ppm	2.3	0.1	19
Sb	ppm	3.7	0.7	9.2
Pb	ppm	322	33	2 667

Source: Environment Agency (1994).

Properties are defined as:

Moisture – the mass loss when a sample is heated to 100 °C until no further mass loss is measured.

Ash – the mass loss when a sample is burned to constant mass.

Gross calorific value – the heat released when a sample is burned and the combustion products are cooled to ambient temperature.

Volatile matter – the loss in mass when a sample is heated to constant mass at 600 °C in a nitrogen atmosphere.

Fixed carbon = 100 – moisture – ash – volatile matter.

SAQ 3

The values for some of the trace components differ by an order of magnitude (a factor of 10). Why might this be the case and how should you treat data such as these?

The UK has a history of cheap landfill and a plentiful supply of sites with suitable hydrogeology. Not surprisingly, landfill has been the predominant method of disposing of wastes (as shown in Table 3). However, with the introduction of recycling and recovery targets and the Landfill Directive, this is beginning to change. Also, in the medium to long term, the planned rises in landfill tax (see the Legislation Supplement for details of the landfill tax rates) will begin to erode the price advantage that landfill has.

Table 3   Waste disposal in England

	Amount produced (million tonnes)	Percentage treated by			
		Landfill	Incineration with energy recovery	Recycling and composting	Other
Municipal					
1997/98	25.7	85	6	8	
2000/01	28.1	78	9	12	
2003/04	29.1	72	9	19	
2004/05	29.6	67	9	23	
2005/06	28.7	62	10	27	
Industrial and commercial (2002/03)	67.9	41	4	42	13
Construction and demolition (2002/03)	90	28	0	50	22

Sources: DEFRA (2005); Environment Agency (2006).

SAQ 4

Why are industrial and commercial recycling rates much higher than those for municipal waste?

1.2.3   Landfill

Landfill is the predominant method of disposing of waste in the UK and many other countries. Although landfill is set to reduce over the years, it will always form a key part of any waste management strategy.

Once landfilled, the organic and biodegradable material present begins to undergo a series of biological and chemical breakdown processes. This was discussed in detail in Section 8 of T210 Block 4: *Wastes management*, but can be summarised as:

> Phase 1: *aerobic degradation* of carbohydrates to sugars, carbon dioxide and water.

> Phase 2: *hydrolysis* of carbohydrates and lipids to form sugars and organic acids respectively. The sugars are, in turn, converted to carbon dioxide and water by *fermentation.*

> Phase 3: breakdown of complex organic acids to simple acids and conversion of the carbohydrates to organic acids, carbon dioxide and hydrogen.

> Phase 4: conversion of organic acids to methane and carbon dioxide, and reaction of hydrogen with carbon dioxide to give further methane.

> Phase 5: oxidation of residual methane to carbon dioxide in the final stages of the process.

Phase 1 takes place in aerobic conditions and only lasts for a matter of weeks, Phases 2–4 are anaerobic with Phase 4 continuing for up to several decades, and aerobic conditions are re-established during the final phase.

These processes lead to the formation of leachate, which can pollute aquifers and surface waters and landfill gas (a mixture of methane and carbon dioxide), which is an explosive hazard and a potent greenhouse gas. Leachate is controlled by fitting the site with an impermeable liner and cap, and installing a pumping system to remove the leachate for treatment. The gas is collected through a network of pipes and pumps and either burned in a flare or used to generate power. The selection of a suitable location for landfill also plays a part in minimising its environmental impact.

Landfills are active for many years after filling has been completed, and active care and maintenance may be required for a century or even longer. For this reason, landfill cannot be considered to be a sustainable solution to waste management. In its favour, landfill treats the entire waste stream and minimises human contact with the waste during processing.

1.2.4   Incineration

Incineration is another well-established waste management option and some European countries treat over half their MSW by this process. The UK treats about 3 million tonnes a year in 15 plants (2005). All MSW incinerators recover energy through power, heat or combined heat and power and the operating conditions and emissions are governed by EU directives. Therefore all incinerators are equipped with gas cleaning systems to remove HCl, SO_2, metals and dioxins. A combination of flue gas recirculation and ammonia or urea injection into the furnace is used to reduce oxides of nitrogen (NO_x) emissions.

Incinerators can either treat unprocessed waste in mass burn or grate furnaces or the waste can be processed by screening and size reduction followed by burning in fluidised bed combustors.

Incineration produces 'bottom ash' as the residue from combustion, which is treated to remove ferrous metal for recycling. The ash is then either landfilled as an industrial waste or used as an aggregate substitute in road building and other construction projects. A smaller amount of residue is also produced by the gas cleaning system which requires processing and/or landfilling as a 'hazardous waste'.

A number of advanced thermal processes are under development or at the pilot or demonstration stage. These processes tend to be based on pyrolysis or gasification. They have some potential advantages over conventional incinerators, but the technology is not yet well established at the commercial scale.

1.2.5   Biological treatment

According to the Waste Strategy (DETR, 2000) 62% of the mass of MSW is biodegradable. This property can be exploited by biological waste management processes. 'Green' biodegradable (garden and kitchen) wastes can be composted in aerobic conditions to produce a soil conditioner, growing media (when blended with other materials) or mulch. The UK as a whole processed 1.5 million tonnes of compostable municipal waste and 0.45 million tonnes of compostable commercial waste at 150 sites in 2003–04 (Slater et al., 2005) and these figures are expected to rise due to the implementation of the Landfill Directive and national recycling targets.

Most composting is carried out in windrows in the open air. However, under the Animal By-products Regulations (HMSO, 2003) kitchen waste must not be composted in open systems (as discussed in Section 4.2.4 below). Also, composting conditions can be controlled more tightly and the negative environmental impacts of the process can be controlled in enclosed systems. For these reasons, it is expected that the amount of composting carried out in buildings or in reactors will increase in the future.

Anaerobic digestion (AD) simulates the processes taking place in landfills in the environment of a controlled reactor. Mechanically processed waste or the separately collected organic fraction of waste reacts in anaerobic conditions to produce a methane-rich gas. The liquid or solid resulting from the process (digestate) has potential use as a soil improver or as a component in growing media. AD is well established for treating farm wastes and about fifty plants globally treat MSW. At present (2006) a number of plants are at the development stage in the UK and two are now operational.

The biological treatment of mixed MSW, known as mechanical biological pretreatment (MBP) is under active consideration in several parts of the UK. The mixed waste is composted, which reduces the moisture content and partially stabilises the waste. This can then be either landfilled with a reduced methane and leachate formation potential or processed to recover metals for recycling, a low-grade biostabilised material for use as landfill cover, and/or a light combustible fraction for incineration.

1.2.6   Recycling

Recycling is the separation of relatively clean components from waste, and their collection and processing to provide new materials. About 22% of the UK's MSW was recycled in 2004–05 (including composting). Whilst this is lower than countries such as the Netherlands, Austria and Germany, where restrictions on landfilling have been in place for several years, the UK has seen a large growth in recycling since the late 1990s and this growth is expected to continue. However, the UK does recycle around 30% of the total amount of controlled waste produced.

Materials for recycling can either be collected from individual households (kerbside schemes) or from centralised banks and skips (bring or drop-off schemes). The former are more expensive, but achieve much better collection rates. In either case, the segregated waste needs processing in a materials recovery facility (MRF) to effect further separation into categories and/or to bulk the material before transport to the reprocessors.

Recycling has a number of environmental advantages in terms of resource and energy saving. However, care must be taken to ensure that these benefits are not exceeded by the vehicle pollution created during collection and transport of the wastes, and the emissions from the MRF and reprocessing plant.

1.2.7   Municipal waste management costs

MSW management is only one of many calls on a local authority's funding, so the cost of providing the service is important. However, much of the waste management activity is carried out in the private sector so costs are often commercially sensitive. As a result, it is not easy to obtain reliable information on the cost of waste collection and disposal. Estimates of the net cost (including landfill tax of £14 per tonne, collection costs and any income from the sale of power and materials) based on Waste Strategy data (DETR, 2000) were presented in T210 Block 4: *Wastes management* and are in the following ranges:

Landfill	£50–55 per tonne
Incineration	£45–70 per tonne
Recycling	£50–140 per tonne
Composting	£90–210 per tonne

These costs are 'internal' or 'private' costs and ignore the 'costs' to the planet as a whole of depleting finite resources and combating the effects of global climate change and other pollutants. Equally, internal costs neglect any benefits from materials and fossil fuel conservation that may result from the waste management activity.

1.2.8   Integrated solid waste management

Integrated solid waste management (ISWM) is a simple concept and is based on the fact that there is no single waste management option that will provide the optimum solution for all components of the MSW stream. An ISWM system could, for example, consist of supplying most of an area with kerbside recycling schemes, collecting organic waste for composting (either through kerbside collections or at civic amenity sites), burning the bulk of the non-recyclable wastes in an incinerator with energy recovery, and landfilling the remaining wastes that cannot be recycled or used for energy recovery.

Such a scheme is being implemented in Hampshire where targets for at least 40% recycling and composting and a maximum of 25% landfill have been set. In this case, the difference of 35% is incinerated with energy recovery.

A key factor in the successful implementation of ISWM is to ensure that the incineration capacity (if required) is designed for the appropriate quantity and composition of the residual waste. Too little capacity and the landfill reduction targets will not be met, whilst too much capacity could jeopardise the success of the recycling activities by diverting potentially recyclable material to incineration.

An ISWM scheme should represent the BPEO for the locality and there are a number of tools such as life cycle assessment (LCA) which allow planners to review all the environmental impacts of different combinations of waste management schemes. These tools are discussed further in Section 8.

2 THE ENVIRONMENTAL AND HEALTH IMPACTS OF SOLID WASTES MANAGEMENT – AN OVERVIEW

The environmental effects of waste management processes range from the extremely localised, such as the noise from a revving collection vehicle in a residential neighbourhood, to the transnational, such as the global climate change impacts of methane escaping from a landfill site. The impacts of these effects cover a similar scale. Most people would regard the sound of a dustcart to be a minor nuisance, whilst the impacts of global climate change could be catastrophic.

When considering the effects of waste management on human health we also see a very wide range of potential impacts from the trivial, such as a short-term odour from a composting site, to the potentially fatal: the effect of smog on people with respiratory or cardiac problems, for example.

When considering environmental impacts it is easy to focus on the negative aspects, but there are positive impacts that need to be considered if we are to present an accurate picture: for example, the recovery of materials and energy and the associated conservation of non-renewable resources.

2.1 Life cycle assessment – an introduction

As I said at the beginning of Section 1, the aim of the waste manager is to collect, treat and/or dispose of wastes while meeting all the regulatory requirements (emission standards, recycling and recovery targets, landfill diversion targets, etc.) and minimising the negative health and environmental impacts. There are a number of sources of guidance and tools to help achieve this that I mentioned in the first section and will discuss again in Section 8. These include:

- waste strategy and related guidance on BPEO
- the waste management hierarchy
- life cycle assessment (LCA) tools.

In particular, LCA can be used to assess different proposed waste management strategies and to help identify the priorities for improving existing waste management systems. This is covered in more detail in Section 8, but the graphs shown in Figure 3 below are based on an LCA assessment of an imaginary area using the IWM-2 software tool (McDougall et al., 2001).

Figure 3 compares the atmospheric emissions from two waste management scenarios. In the first, all the waste goes to a landfill fitted with a gas collection and recovery system that generates electrical power for the grid. In the second case, an integrated solid waste management (ISWM) scheme has been implemented. In this scheme, 28% of the waste is recycled or composted, 58% is incinerated and the remaining 18% is landfilled.

The first point to note is that several of the emissions are negative. This recognises the fact that recycling and energy recovery reduce the environmental impacts due to the extraction and processing of virgin raw materials. So the negative impact of sulphur dioxide in the ISWM case represents a reduction in SO_2 emissions because the incinerator reduces the amount of coal-fired electricity generation required. These savings are often referred to as 'offsets'.

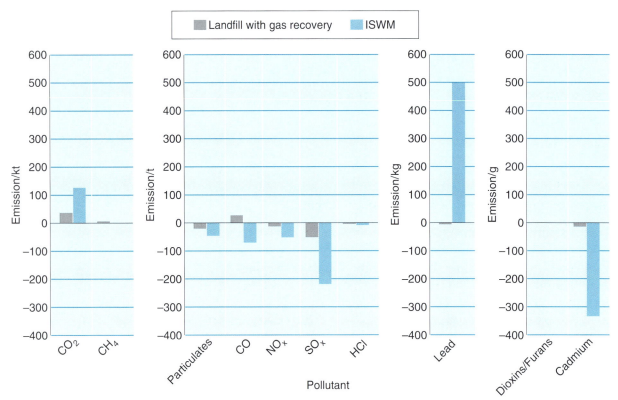

Figure 3   LCA comparison of landfill and ISWM

It is important to note that the different emissions are all in different units and they simply represent mass emissions rather than environmental impacts. Therefore, comparisons should not be made between columns. You should also note that the size of an emission does not necessarily bear any relationship to its environmental impact. Assessing this relationship requires the user to find a way of comparing and aggregating different impacts. It is relatively simple to compare the global warming potential of carbon dioxide and methane, but it becomes much more difficult to compare (say) X tonnes of the greenhouse gas carbon dioxide with Y cubic metres of landfill leachate. It is easy to see why any such evaluations are fraught with difficulties. However, as you will see in Section 8, some of the more detailed LCA packages do go some way to making such comparisons.

Having said this, it is possible to use LCAs to compare two processes or options, which helps to evaluate their respective impacts and also to identify the priority areas if the impacts are to be reduced. Looking at Figure 3 we can conclude that, with the exception of carbon dioxide and lead emissions, ISWM produces the lowest emissions. Looking deeper into the source data we find that:

■ the CO_2 emissions from ISWM are due to the incineration option

■ the lead emissions from ISWM are due to the recycling option.

This immediately tells us that improvements in the emissions from the ISWM scenario could be made if the amount of energy recovered from the incinerator per tonne of waste burned was increased. For example, if the incinerator was part of a combined heat and power scheme rather than a power-only system, the overall thermal efficiency would increase from around 20% to up to 80%. This would give up to a fourfold increase in the fossil fuel saved and therefore an increase in the CO_2 offset and a reduction in the size of the CO_2 bar on the graph.

The reason for the high lead emission from the recycling process is unclear, but it does indicate that further research should be carried out in order to identify the source of this atmospheric pollutant.

Figure 3 covers releases to only one medium. It is possible to draw up a series of tables looking at different impacts. The IWM-2 package that I used to produce the figure also allows the user to compare:

■ fuels consumed and saved

■ final solid waste landfilled (hazardous and non-hazardous wastes)

■ emissions to air

■ emissions to water (ground and surface water combined).

These all need to be taken into account along with the financial costs when evaluating different waste management systems.

Life cycle assessment goes some way to providing an objective way of comparing the environmental impacts of different waste management options. However, it has done little to reduce the controversy when it comes to selecting options. It is not unknown for the proponents of two different schemes or technologies to use the same LCA data to arrive at completely different conclusions. This is something you will meet in Exercise 3 in Section 5.

2.2 Emissions and impacts of waste management options

Tables 4–8 below provide a brief qualitative overview of the main sources of emissions and potential health impacts of different waste management operations, the emissions from each source and the potential environmental and health impacts.

Table 4   Emissions and impacts of waste collection and transport

Source	Emission/effect	Potential health impact	Potential environmental impact
Vehicle engine emissions	CO, NO_x, CO_2, diesel exhaust particulates, e.g. PM_{10}, volatile organic compounds (VOCs)	Respiratory and cardiac problems in sensitive subjects (from respirable and inhalable dusts)	Greenhouse gases, smog, deposition of diesel exhaust particulates
Vehicle non-engine emissions	Rubber dust, particulates		
Vehicle fuel emissions	VOCs		
Collection	Nuisance dust, noise, odours, bioaerosols, road traffic accidents	Nuisance, injury, death	Noise nuisance, general increases in road traffic
Manual handling		Musculoskeletal injury, cuts/skin penetration, crushing injuries, slips, trips and falls	
Vehicle and bin cleaning	Liquid wastes to sewer (surfactants, BOD, COD)		Potential for ground and surface water contamination

Table 5   Emissions and impacts of MRF operation

Source	Emission/effect	Potential health impact	Potential environmental impact
MRF operation	Noise, engine emissions including diesel exhaust particulates, PM_{10}, bioaerosols, odours	Worker respiratory problems, hearing issues if noise excessive	
	Bending, lifting, injuries from broken glass, etc.	Musculoskeletal injury, repetitive strain injury, cuts/ skin penetration	
Fugitive emissions	Worker/neighbour exposure to nuisance dusts, respirable dust, PM_{10}, bioaerosols, VOCs, noise, odours	Noise and odour nuisances, possible respiratory problems in sensitive subjects	Nuisance issues
Materials reclaimed			Conservation of resources

Table 6   Emissions and impacts of composting

Source	Emission/effect	Potential health impact	Potential environmental impact
Shredding of feedstock	Noise, engine emissions, PM_{10}, bioaerosols	Noise and odour nuisances, possible respiratory problems in sensitive subjects	Transport of various species of micro-organisms downwind, dust affecting photosynthesis of surrounding vegetation
Composting (including pile turning)	Noise, engine emissions, PM_{10}, bioaerosols, CO_2, ammonia, VOCs	Noise and odour nuisances, possible respiratory problems in sensitive subjects	Contribution to global warming, possible localised nitrate deposition
Screening	Noise, engine emissions, PM_{10} and potentially ultrafines ($PM_{2.5}$ and below), bioaerosols	Noise and odour nuisances, possible respiratory problems in sensitive subjects	Transport of various species of micro-organisms downwind, dust affecting photosynthesis of surrounding vegetation
Fugitive emissions	Worker/neighbour exposure to dust, bioaerosols, VOCs, ammonia, noise, odours	Noise and odour nuisances, possible respiratory problems in sensitive subjects	Nitrification of surrounding soils
Compost product		Address with standards such as PAS 100	Peat bed conservation, reduction in landfilling of biodegradable wastes

Table 7   Emissions and impacts of incineration

Source	Emission/effect	Potential health impacts	Potential environmental impact
Flue gas emissions	CO_2, NO_x, HCl, PM_{10}, SO_2, metals, dioxins	Respiratory problems, cancer	Contribution to smog, acid precipitation, global warming
Solid residue management	Leaching of metals	Drinking water contamination	Groundwater and surface water contamination, reduction in landfilling of biodegradable wastes
Fugitive emissions	Worker/neighbour exposure to respirable dust, bioaerosols, VOCs, noise, odours	Possible respiratory problems in sensitive subjects	
Thermal output			Fossil fuel savings (and possible associated emission reductions)

Table 8   Emissions and impacts of waste landfill

Source	Emission/effect	Potential health impacts	Potential environmental impact
Leachate escape	BOD, COD, metals, organic solvents	Drinking water contamination	Surface water and groundwater contamination
Treated leachate discharge	Controlled discharge of the above	Drinking water contamination	Surface water and groundwater contamination
Landfill gas fugitive emissions	CH_4, CO_2, odours, VOCs	Nuisance, injury/death from explosions	Damage to plants, contribution to global climate change
Landfill gas engine/flare emissions	NO_x, CO, VOCs, SO_2, PM_{10}, CO_2, dioxins	Nuisance, respiratory problems, cancer	Contribution to smog, contribution to global climate change
Thermal output			Fossil fuel savings (and possible associated emission reductions)
Fugitive emissions	Worker/neighbour exposure to PM_{10} bioaerosols, VOCs, noise, odours	Noise and nuisance, possible respiratory problems in sensitive subjects	

PM_{10} is discussed in Block 2: *Managing air quality.*

Bioaerosols and their impacts are discussed in Section 3.2.2.

SAQ 5

For Tables 6 and 7 (composting and incineration) draw up a list of what you would expect to be the main health and environmental impacts arising from each technology.

You have probably deduced that the main impacts relating to waste collection, MRF operation and composting tend to be localised and health-related. All the emissions are fugitive in nature, they are released close to ground level and, with the exception of the gaseous emissions from composting, relate to dust, various types of particulates, odours and noise. These localised emissions tend to be small in magnitude, but can potentially lead to significantly high concentrations in the vicinity of the discharge, affecting the site, the workforce and those living or working close to the site.

In contrast, the main potential emissions from landfill are leachate and landfill gas, which tend to have environmental impacts at the regional, national and even transnational levels. Similarly, the main impacts from incineration are due to the high level of gaseous discharges. As you will see in Sections 5 and 6, the principal impacts of such emissions relate to environmental rather than to health issues.

This is reflected in the following sections where the emphasis is on:

- waste collection: health and safety, micro-organisms, localised traffic issues
- MRF operation: health and safety and occupational exposure to dust, micro-organisms and noise
- composting: health and safety, occupational and local exposure to dust, bioaerosols and odours and, to a lesser extent, process emissions
- incineration: gaseous emissions, solid residue management
- landfill: landfill gas, leachate.

Recycling also has impacts when the reclaimed materials are reincorporated into new materials (for example when glass is re-melted and blended with virgin glass). These impacts can be both positive (such as the reduction in energy use when manufacturing aluminium from scrap rather than from raw bauxite ore) and negative (such as the production and landfill of contaminated sludge from the de-inking of recovered paper). When assessing the overall impact of waste management options, these impacts are important so are taken into account in LCA assessment packages. You need to be aware of these potential impacts, but they are not covered as part of this course. It can be argued that once a reclaimed material enters a manufacturing process, it ceases to be a waste at that point and this is reflected in the national regulatory regimes. For example, the Aylesford paper mill, which reprocesses around 470 000 tonnes of reclaimed newsprint a year, is regulated under IPPC as a papermaking plant rather than as a waste disposal facility.

3   ENVIRONMENTAL IMPACTS OF WASTE COLLECTION AND TRANSPORT ACTIVITIES

3.1   Introduction

This section and Section 4 deal with the parts of the waste management cycle where the environmental impacts tend to be more localised and, regrettably, often neglected in the 'waste management debate'. The activities covered in these chapters – waste collection, recycling and composting – all need to be considered from the perspectives of resource and energy conservation and major pollutant releases, but the local impacts such as dust, volatile organic compounds and odours are far more important from the point of view of the people carrying out the activities and the local communities.

Much of the following sections are concerned with occupational exposures to dust, heavy metals and bioaerosols (defined in Box 2 below). The Health and Safety Executive (HSE) publishes an annual list of workplace exposure limits (WELs) in its EH40 guidance document (HSE, 2005). A WEL can either be based on a longer term 8-hour time weighted average (TWA) or a shorter exposure time over 15 minutes. WELs replaced the old system of occupational exposure limits (OELs) and maximum exposure limits (MELs) in 2005. A selection of WELs (current in 2005) is given in Table 9.

Table 9   Workplace exposure limits (mg m^{-3})

Substance	8-hour TWA (mg m^{-3})	15-minute period (mg m^{-3})
Total inhalable dust	10	
Respirable dust	4	
Cadmium and cadmium compounds[*]	0.025	
Mercury and mercury compounds	0.025	
Hydrogen Sulphide	7	14
Ammonium chloride fume	10	20

Source: Data from HSE (2005).

[*]excluding cadmium oxide fume and calcium sulphide

See the T308 Air block and T210 Block 2: *Managing air quality* if you need to remind yourself of the differences between inhalable and respirable dusts.

Not all substances are listed in EH40 but may still have a WEL. For instance, lead is not listed as it has separate legislation (Control of Lead Regulations 2002), but still has a WEL (0.15 mg m^{-3} as an 8-hour TWA).

The introduction of WELs further defined 'good practice' with the introduction of eight guiding principles to workplace exposure, which apply regardless of whether there is an occupational exposure limit of any kind. These include issues such as designing a process to minimise emissions and so exposure; taking into account all relevant exposure routes, e.g. inhalation, dermal, ingestion; controlling exposure proportionate to the health risk; informing and training employees etc. This means adequate control of exposure needs to take into account the guiding principles, ensure a relevant WEL is not exceeded and ensure any exposures are reduced to as low as is reasonably practicable.

WEL guidance is intended to feed into COSHH regulations (control of substances hazardous to health).

The situation over exposure to the fine solid or liquid particles that contain micro-organisms (known as bioaerosols) is more complicated. For bioaerosols, the Environment Agency adopts a risk-based approach to allowable concentrations. This is further reviewed in Section 3.2.2 below.

3.2   Waste collection

For as long as we have produced waste, there has been a need to collect and dispose of it, and evidence for the existence of landfills goes back to 3000 BC. In the UK, the Public Health Acts of 1848 and 1875 were the first which required local authorities to collect waste at weekly intervals from households, specifying a 'movable receptacle' in which waste was placed for collection. The first collections for recycling included waste paper collections and 'rag-and-bone men' (a brief history of waste collection is given in Chapter 1 of T210 Block 4: *Wastes management*). However, for much of the 1900s waste collection was a case of emptying metal dustbins into a lorry and, more recently, simply collecting and loading black plastic bags. In the latter half of the century the lorries began to be replaced by purpose-designed refuse collection vehicles (RCV) which compacted the waste, increasing the vehicle payload. Since the 1960s, mechanisation of collection and disposal has increased – effectively separating waste from operatives. In the 1990s wheeled bins came to the fore; primarily these were sold as a cleaner, more hygienic and safer way to collect waste from households. It enclosed the waste, separated it from the collector and delivered it to the RCV more efficiently and safely than black bags (which run the risk of splitting and can cause injuries to collectors from potentially sharp objects). The use of wheeled bins also remove the need to lift the containers manually, an important factor I refer to in Section 3.2.2 below. The use of wheeled bins is still increasing for these reasons and the fact that the strong rigid container makes fortnightly collections more acceptable to householders.

The project material included on the Resources DVD shows a video clip of wheeled bin waste collection and summarises the various advantages of this method.

However, the main disadvantage of introducing wheeled bins is that it can encourage householders to produce more waste as they take advantage of the closed receptacle and deposit materials that they would never have put in a plastic sack (in particular, garden waste, DIY waste and other building materials). This can make wheeled bins extremely heavy to manoeuvre, and the measured increase in waste production can affect a local authority's recycling performance – this emphasises the importance of fully evaluating a collection system prior to introduction.

As mentioned above, waste collection used to be uniform across the country – black bags and compactor trucks. However, this is no longer so. The introduction of wheeled bins and increasing need to meet recycling targets has been the main driver in changing this; collection systems are no longer standard and they are continuously undergoing further change due to the varied requirements for materials to be segregated at the kerbside. This has had a major impact on the way waste is collected.

Kerbside collection schemes for recycling are also outlined in Section 11 of T210 Block 4. These provide various types of containers (boxes, bags, wheeled bins, etc.) for the collection of different components from the waste stream. Some recyclables collections are carried out in conjunction with the 'normal' waste collection, whilst others take place on a different day or at a different frequency (perhaps fortnightly rather than weekly). The range of materials collected also varies widely. Typical examples of the types of receptacles and materials commonly chosen are outlined below:

- kerbside sorted box or bag collection: either for the collection of a single material such as paper, or co-mingled 'dry recyclables' e.g. paper, card, cans, plastics and occasionally glass, normally in conjunction with a general waste collection for the remainder

- kerbside bag collection with coloured sacks: one for 'dry recyclables' and one for the remainder (with separation and sorting taking place at the MRF)

- twin wheeled bins: one for dry recyclables and one for the remainder

- a variety of wheeled bins: e.g. one for dry recyclables, one for organic wastes and one for the remainder

- separate collection of organics via wheeled bins or in some cases biodegradable bags.

The vehicle used also varies greatly with different collection schemes. With some collection schemes co-mingled recyclables are further sorted at the kerbside into different bins on a specially built recycling vehicle. Box-based schemes often use this approach.

Other types of vehicles are divided longitudinally into two compartments; materials for recycling are carried in one compartment and residual waste in the other. This system is often used in conjunction with wheeled bin collections and has the advantage that the number of collection vehicles passing each house is reduced (with a corresponding reduction in the environmental impact of the collection).

Some recycling schemes use a conventional lorry with a cage on the rear in which the materials are emptied, with the residual waste collected using a separate standard RCV. This has lower set-up costs in terms of vehicle costs, but labour costs for both collecting and final sorting of the materials can be higher.

Despite the plethora of alternatives available when choosing a collection scheme there are also some issues that remain constant whichever scheme is selected, i.e. it needs to be cost effective: collecting the greatest amount of material in the least amount of time for the lowest possible cost. This is normally measured either by unit cost per tonne or cost per household. In order to reduce costs local authorities may choose to:

- increase the number of households served by increasing the number of households from which each vehicle collects

- increase the quantity of recyclables collected from households already served, e.g. increase the participation rate, or

- increase the effectiveness of the separation so that an increased quantity is collected in each vehicle.

In addition, some authorities reduce the collection frequency of the non-recyclable waste to once a fortnight. This reduces collection costs and also encourages householders to participate in the recycling scheme and produce less residual waste.

The Resources DVD project material shows the operation of kerbside recycling collections. One box-based scheme collects glass and newspaper one week, and cans and plastics the next week. The contents of the boxes are sorted at the kerbside into a specialised collection vehicle for the recycling scheme. The other scheme shown involves the collection of mixed recyclables (excluding glass) in plastic sacks with all the sorting taking place at the MRF.

There are many factors that affect the efficiency of a collection scheme, helpfully outlined by Hummel (2004) and summarised in Box 1.

BOX 1   FACTORS INFLUENCING THE EFFICIENCY OF KERBSIDE COLLECTION SCHEMES

Operational factors

- type of collection, e.g. number of kerbside sorts
- vehicle capacity (weight and volume)
- productive time available for collection: total hours worked each day
- non-productive time, e.g. unloading, driving to/from collection round
- size of collection crew
- physical properties of targeted materials (influences ease and speed of collection and weight/volume characteristics of vehicle)
- collection/container method used and how easy it is for the crew to use.

Household factors

- numbers participating in the scheme
- participation frequency, e.g. each collection or infrequently
- quality of participation
- quantity of targeted materials correctly separated and set out (capture)
- quantity of other materials incorrectly put in containers (contamination)
- how well the householders use the system, e.g. using containers provided.

Waste auditing looks at a number of factors in a kerbside collection scheme. It examines where the waste arises, what sort of materials are produced by the households (which could mean hand sorting the contents of a representative selection of waste containers), where it goes to and how much it costs to do this.

In order to determine these factors, local authorities might measure performance with participation surveys and waste auditing.

However, due to the variety of variables it is very difficult to make direct comparisons of different schemes, as noted by Hummel (2004). What might be a good scheme for one area might not necessarily translate into a good scheme for another; hence the wide range of different methods used.

It is also important to note that separate collection and recycling/composting schemes for household waste should not be viewed in isolation; they should form part of an integrated waste management programme. For example, the collection of paper for recycling would reduce the calorific value of the remaining waste, while the collection of kitchen and garden waste would increase the calorific value. This could be important if the residual waste is to be processed by incineration or other thermal processes. The amount of residual waste remaining must also be taken into account when planning its disposal, be it by thermal processing, other treatment or landfill.

3.2.1   The role of the householder

For more traditional methods of waste collection the householder is required to do very little: simply put out the bags or bin on the same day of each week. With separate collections for recycling and/or composting schemes the willingness of the householder to participate is the first important link in the chain and determines whether the scheme will be successful or not. This will depend on how easy and convenient it is for the householder to separate the required materials and on the quality and distribution of the instructions and publicity supplied by the scheme operators. Again, it must be stressed that no single type of collection scheme is the 'best', as different people, housing and areas will all have schemes that suit them better than in other situations.

However, Hummel (2004) outlines a number of key factors which can influence householder participation:

- collection frequency: has to be easy to remember (every Monday or alternate Wednesdays rather than the third Thursday in every month beginning with 'M' and Tuesdays in months beginning with 'J', etc.)

- complexity of household sorting: clear instructions for which materials are to go into which container are needed, otherwise people can get exasperated (this is understandable in schemes which collect, e.g. yogurt pots but not margarine tubs)

- materials targeted: need to be recognisable; mistakes are often made with plastics. Householders find it difficult to differentiate between different polymer types so tend to put all plastics in their containers; very clear instructions are needed to make sure that householders only collect the correct types.

- type of container: for instance, wheeled bins for flat dwellers are not a good idea

- educational material: needs to be timely, i.e. not during the holidays, and should be reinforced – a one-off leaflet can be lost and householders can forget instructions easily

- changes to the collection: if changes are not minimised householders can get confused

- expansion to the collection: if it affects existing participants they might lose interest.

The most important point Hummel (2004) makes is that 'collecting 80% of materials from 20% of households will be much more cost effective than collecting 20% of materials from 80% of households, and the overall quantity will be the same'. This point illustrates how important the role of householders is. Currently, they can be asked to participate, but they cannot generally be made to participate, and any recycling scheme is very likely to be asking for a lot of extra effort over traditional waste collection. Therefore the key is to make it easy.

There are exceptions to voluntary participation, however. The London Borough of Barnet was the first to introduce a compulsory recycling scheme in April 2004. They provided a box for dry recyclables and a wheeled bin for the residual waste. Residents who fail to recycle are first visited by recycling assistants, and subsequently are issued with warnings and formal notices. As a last resort the Council reported that it would prosecute persistent offenders (London Borough of Barnet, 2004).

However, it is important not to forget the requirements and impacts on the householder. The main issues are: the type of container(s) (position, size and numbers); space for storage (ventilation and hygiene, prevention of access by vermin, lighting, security, construction materials for store, size, access and marking); the transportation of material usually to the kerbside and the types of material collected (dry recyclables, biodegradable and/or hazardous wastes).

In the Netherlands these issues have been investigated in more detail. Wouters et al. (2000) investigated the association between indoor storage of organic waste and concentrations of micro-organisms in the home (these are discussed in more detail in Section 3.2.2 below). This study found that indoor (usually kitchen) storage of separated organic materials significantly increased the concentrations of micro-organisms in dust samples – but more importantly this was associated with health effects in asthmatic children. Hence it really is important to consider impacts on householders when designing an organic waste collection programme, particularly if fortnightly collections are proposed.

3.2.2   Health and safety issues associated with collection

From a health and safety viewpoint, waste collection is not regarded as a 'site' as such, but the Health and Safety Executive (HSE) does regard it as a workplace. Obviously collection of waste involves handling of bags or containers of some kind. Wheeled bins replace lifting and carrying of heavy bags, but could lead to a more intensive repetitive work regime placing heavier workloads on arms and wrists; weight and unevenness of road surfaces can also cause problems (GMB, 2003). In 1993 the Danish Working Environment Service issued a new set of regulations concerning collectors, ensuring waste should not be carried or loaded manually onto collection vehicles; limiting pushing and pulling forces (200 Newtons or 20 kg constant movement, 400 Newtons for short jerk to begin load moving), and specifying hygienic requirements, particularly for collection of source-separated organic waste (Nilsson, 1996).

Indeed, the Health and Safety Executive (HSE) in the UK has also become concerned about pushing and pulling forces (particularly an issue with wheeled bins), and recently commissioned a study examining this (Ferreira et al., 2004). The authors reported that HSE's own accident database revealed that 11% of manual handling injuries between 1986 and 1999 were related to pushing and pulling, not only physical overexertion, but also limb trapping, slipping and falling, equipment breakages and conditions of the work environment – all important issues for waste collectors. As a result revised guidance advises that where critical risk factors such as uneven floors, confined spaces, kerbs, ramps and slopes, and trapping hazards are present, a detailed pushing and pulling risk assessment should be undertaken. This research recommended a checklist that should be carried out before a load is handled, and acknowledged that a number of factors could influence how a load is handled, including design features such as: the type of container, dimensions, loading factors, handle characteristics, wheel and castor characteristics, conditions of the work environment and maintenance. As a result of the research, pushing and pulling risk filter guidelines for starting and stopping a load were reduced to 20 kg for men and 15 kg for women (assuming the maximum travel distance is no more than 20 metres). This value would often be exceeded in the case of a collection round with wheeled bins.

In terms of physical injuries, areas where collectors are most at risk include: back, shoulder and arm injuries; sharp objects (e.g. broken glass) and hypodermic needles; violence from members of the public (usually traffic related); pets of the households served; and the RCV (when reversing) and other traffic (GMB, 2003). A recent HSE-funded report (Bomel, 2004) reported that the greatest risks to collection staff included falls from height (climbing from the cab of the vehicle), traffic accidents, kerbs and footpaths leading to tripping accidents and lifting or moving heavy containers (including throwing black bags into an RCV). Wheeled bins were specifically mentioned as sometimes being of 'excessive size'. Other highlighted issues included noise from glass breakage and violence towards staff from the public. Indeed, the Bomel report (2004) found that waste management had one of the highest accident rates for all sectors; the overall accident rate for the waste industry in 2001–02 was estimated to be around 2500 per 100 000 workers. This is around four times the national rate (559 per 100 000 workers as reported by HSE); the rate of fatalities was 10 per 100 000 (0.9 national rate) and major injuries 330 per 100 000 (101 national rate).

Items which break the skin could also possibly be contaminated by a life-threatening risk, such as tetanus or hepatitis. These types of hazards are prevalent when collecting waste, but may also have implications in disposal and recycling activities. Working with plastic sacks has also been linked with eczema (Malmros, 1990).

It is important to note that there are many potentially hazardous items in household waste. These can include: disposable nappies and sanitary protection items, medical items, paint and related products, herbicides, pesticides and fungicides, motoring products, household cleaners, water treatment chemicals, adhesives and glues, photographic chemicals, aerosols, gas cylinders, fluorescent tubes, batteries, smoke alarms, CFCs and other chemicals. Collectively, these materials are known as 'household hazardous waste' (HHW). The periodic clearing out of garden sheds and garages is a major source of HHW. A survey by Kerrell (1991) found HHW in the UK contained items such as old lead-based paints, chemicals and medicines. Some of these had been stored by householders for up to 25 years, and some labels were missing or were too dirty to read. Chemicals found included carcinogens, poisons, toxins, corrosives, mutagens, teratogens, inflammables and those known as 'persistent within the environment'.

These are all perfectly legitimate items and, while many local authorities encourage householders to bring household chemicals, oils, paints, batteries, etc. to collection points at civic amenity sites, in practice a householder can legally dispose of almost anything with their household waste.

In addition to the physical risk of handling these items, HHW disposed of in household waste can cause equipment and property damage, most commonly fires. These result from either a flammable substance coming into contact with an ignition source, or different materials from different sources mixing and causing a reaction. Compactor vehicles such as RCVs are at particular risk, as they liberate substances from their containers on crushing. Aside from the risk to householders storing these products, both vapours and liquid remnants could cause hazards to collectors. At present, much HHW is disposed of incorrectly, or stored for long periods of time. These storage and disposal practices are under review in many European countries, and separate collection of hazardous wastes is practised in many of these.

Additionally there is a lot of concern over the exposure of workers during waste collection to dust and bioaerosols containing micro-organisms of bacteria and fungi (which in turn contain endotoxin and glucan). These terms are defined in Box 2.

BOX 2   MICRO-ORGANISM TERMINOLOGY

Dust: particulate matter that may be either suspended in the air for long periods of time or deposited.

Bioaerosols: airborne solid or liquid particles suspended in air, which may contain micro-organisms, ranging in size from 0.5 to 100+ μm, hence are invisible to the naked eye (an aerosol of very tiny biological particles).

Bacteria (singular, bacterium): a group of micro-organisms with a primitive cellular structure, in which the hereditary genetic material is not retained within an internal membrane (nucleus).

Endotoxin: part of the cell wall of some bacteria which is liberated during cell division or when a cell dies.

Fungi (singular, fungus): a group of micro-organisms with a more complex cellular structure than bacteria, in which the hereditary genetic material is retained within an internal membrane, forming a nucleus. Often referred to as mould.

Glucan: part of the cell wall of fungi.

Micro-organisms: microscopic organisms that are capable of living on their own. Often simply called 'microbes'. Micro-organism concentrations are quoted in terms of colony-forming units per cubic metre (cfu m^{-3}) determined by their ability to grow when cultured in the laboratory.

Pathogens: generally viable particles with the potential to infect.

Where do these airborne contaminants originate? Quite simply, in any biodegradable components of the waste stream. That means in any food and garden waste you throw away, but also remnants in food containers, and even paper and cardboard if it gets wet. As soon as waste is disposed of it starts to break down. If the weather is warm, this could happen very quickly, and very often the material also dries out. Once dry it becomes easily aerosolised as a dust containing micro-organisms, and it is this dust that can be a problem if inhaled (recall respirable dusts in Block 2: *Managing air quality*). Bioaerosols can affect health in many ways, depending on the exposure route.

SAQ 6

What are the potential exposure routes for bioaerosols?

For each route of exposure a number of factors will need to be considered, e.g. whether effects are likely to be acute (short term) or chronic (long term), whether there is a risk of sensitisation (of respiratory tract if inhaled, or skin if contact occurs) or allergic reaction, whether it could be harmful to the reproductive process or whether infection could be caused (such as in the case of micro-organisms). However, the main routes for bioaerosol exposure are generally inhalation and ingestion.

In Denmark, Nielsen et al. (1995) examined waste collectors and found exposure depended on what tasks they were doing. 'Loaders' onto the truck, followed by 'runners' collecting bins, had the highest contact with waste and higher exposures than the driver who spent most time in the vehicle. Most bioaerosols were thought to be generated when waste was put into the vehicle or compressed. This was demonstrated, as samples on the roof of the truck were significantly higher by a factor of 10 than samples taken elsewhere on the vehicle. The exposures were thought sufficiently high that this could have an implication in illness and health complaints expected.

In order to determine ways in which to reduce exposure of dust and micro-organisms to waste collectors, Nielsen et al. (1997) examined three different types of RCV (a basic RCV, one with a plastic curtain and another with a plastic curtain and exhaust air) to determine the amount of dust liberated during compaction, both in the field and in the laboratory using wood dust. The basic RCV gave significantly higher exposures than the RCV with a plastic curtain and exhaust air. It was also noted that in the laboratory quick emptying (speeding up the process) with the curtain and exhaust RCV caused a significantly lower dust concentration than the other options. Slow emptying of the RCV with curtain and exhaust gave a significantly lower dust concentration than the basic RCV. When a top-loading RCV (4 m high) was compared with an RCV which loaded closer to the ground (1.5 m high) exposure to dust and micro-organisms for collectors was also lower with the top loader. This study illustrates how the design of the truck can affect potential worker exposure.

Nielsen et al. (1997) also investigated the separate collection of garden waste and collection staff exposure. Around 75% of RCV operators engaged in garden waste collections were shown to be exposed to high levels of bioaerosols, particularly from exposure to fungi. As a comparison, only 23–38% of mixed waste collectors and fewer than 20% of other types of waste collectors were similarly exposed.

The Norwegian Pollution Control Authority (SFT) also studied source separation of household waste and impact on health and working conditions of collectors (Heldal, 1997). This research compared closed-wheeled bins, an aerated container and a source sorting cabinet with four boxes in different parts of Norway, using low-loading and high-loading RCVs. It was found that there was a significantly higher exposure during RCVs loading 1 m above ground, compared with 2 m, but only in summer. Residual fractions (i.e. remaining waste) collected every four weeks showed slightly higher exposure levels than garden waste. In general, the proportion of fungal spores was higher during the summertime. It was concluded that potential exposure levels during collection of waste were relatively high, and decreased by up to a factor of two in winter. The study concluded that collected garden waste showed higher exposure levels than unsorted waste, and closed bins for collected garden waste were considered to produce higher exposure levels.

In terms of health studies Malmros et al. (1993) studied a collection system in Faaborg (2549 households) and Herning (1274 households) during 1990–91. This study concluded there were no significant differences in micro-organisms in clean or dirty containers (wheeled bins), whether the waste was 5–14 days old, or by season. It also concluded that the amount of liquid produced by the waste biodegrading increases with time, with bioaerosols spread on emptying. Because of this it was recommended that lids should open automatically to limit operatives' exposure and the release of any liquid produced should be prevented during compaction. The report also studied 26 of the 31 operatives collecting 'green' waste with different collection systems (6 from Faaborg and Herning), and reported that 11 of the 26 operatives had summer symptoms of nausea, diarrhoea, general malaise, tiredness and fever (these are typical symptoms after exposure to bioaerosols).

Ivens et al. (1999) investigated 1747 waste collectors, and found an exposure-response relationship between nausea and endotoxin exposure and between diarrhoea and exposure to both endotoxins and viable fungi. In both cases, lower exposures were related to a lower number of reports for nausea and diarrhoea. This study is important as it illustrates that exposure to bioaerosols is linked to certain effects on the health of collectors.

The problem with these studies is that we have no set exposure concentrations, unlike for dust, at which we know bioaerosols have an effect on us – indeed it is thought that different concentrations may affect different people in different ways. Hence there is little idea on what constitutes 'safe' levels of bioaerosols to which an individual can be exposed with respect to waste management, either in terms of occupational exposures or possible environmental risks. What these studies do is emphasise the microbiological potential of what are known as 'organic dusts', which are mixtures of widely varying constituents.

The handling of waste is associated with a wide range of respiratory diseases and symptoms in exposed individuals, such as dry cough and asthma type symptoms, probably attributable to high levels of micro-organisms containing a mixture of fungal spores, bacterial endotoxins and allergenic proteins with toxic and immunological influence. The exact mechanisms involved and the components or combination of components of the dust that elicit specific effects are not well known, and dose responses are not well characterised. This means bioaerosols have no universally accepted 'dose-response' relationship. This is a quantitative relationship, a correlation between the dose of a micro-organism and an effect caused by the micro-organism (dose-response relationships are discussed in T210 Block 1: *The environment, risk and public health*). At present there is still some uncertainty over this relationship regarding bioaerosols.

Therefore, related legislation and guidance, from reviews, consensus and research from a variety of workers are used as guidelines based on observed effects (it should be noted many studies looked at generally uncontaminated indoor environments), as seen in Table 10.

Table 10   Recommended exposure limits

Source	Recommended limit	Potential health effects
Lavoie and Guertin (1991) (USA)	10^4 cfu (colony-forming units) m^{-3}	10^{6-8} and above known to cause allergic alveolitis. Other health effects depend on particular micro-organisms present
Breum et al. (1997) (DK) Rao et al. (1996) (various standards reviewed)	10^2–10^3 cfu m^{-3} (non-contaminated indoor environment)	
Environment Agency (2003a)	10^3 cfu m^{-3} bacteria and fungi 10^2 cfu m^{-3} gram-negative bacteria	gram-negatives linked to endotoxins

SAQ 7

Think back to Block 2: *Managing air quality*. What complications might be present in setting occupational exposure limits for micro-organisms?

Very little research has been carried out on what constitutes 'safe' levels of bioaerosols to which an individual can be exposed with respect to biodegradable wastes, in terms of either occupational or environmental exposures. Hence there are no recognised exposure limits for bioaerosols, and little dose-response data. This general lack of data also means it is very difficult to draw on past studies and provide definitive conclusions. Therefore, related legislation and guidance, and research from a variety of workers are used as a guideline for exposure.

Lavoie and Guertin (1991) suggested 10^4 cfu m^{-3} (colony-forming units per m^3 of air sampled) as an occupational guidance level for waste management facilities based on work by a variety of authors, but this is an approach based on a review of the literature and is not supported by dose-response data. The Danish Working Environment Service proposed that levels in excess of 10^6–10^9 cfu m^{-3} could cause respiratory problems (Wurtz, 1996). Lacey and Dutkiewicz (1994) report that concentrations of viable micro-organisms above 10^6 cfu m^{-3} have been linked to hypersensitivity pneumonitis (allergic alveolitis) complaints, e.g. 'farmers' lung'. Wheeler et al. (2001) use a more conservative concentration of 10^3 cfu m^{-3} for bacteria and fungi, effectively a no-observed effect level (a concentration at which it is expected that there is no risk to the majority of people). However, given that natural concentrations could be regularly expected to exceed this, and that farming regularly generates concentrations in excess of this (Swan et al., 2003) and again it is not supported by dose-response data, this can serve as a guideline only.

In conclusion, there are a number of health and safety issues concerned with waste collection. Poulsen et al. (1995b) summarised some of the main issues as outlined in Table 11.

Table 11   Summary of hazards to traditional waste collectors

Hazards for collectors	Exposures	Health issues which could result
Bacteria, Gram-negatives, fungi, endotoxin, glucan	High (10^6) exposures to bioaerosols and endotoxin	Pulmonary, gastrointestinal, mucous membrane and skin problems (worse during warm months)
Diesel fumes	Unknown	Increased risk of pulmonary disorders, but little data
Hazardous household waste	Unknown	Potential fumes, metals, increased sick leave

It can be concluded that risks of exposure to dust and bioaerosols in waste collection are high. Poulsen et al. (1995b) concluded that incident rates and causality of problems are sparsely reported and there are little data to substantiate the range and extent of these hazards. They found that waste collectors are exposed to a complex mixture of bioaerosols and airborne volatile compounds. Synergistic interactions between exposure agents were also thought to be important when studying causality of health problems among waste collectors: for example, diesel fumes may increase action of allergens in susceptible individuals (Poulsen et al., 1995b).

3.2.3   Environmental issues associated with collection

The main environmental issues associated with collection are noise and vehicle exhaust emissions (which link with the earlier blocks on the subject). The main noise impacts are a nuisance rather than a hazard: for example, vehicle engines idling and revving, noise from any compaction operations and the sound of glass being dropped in metal compartments during recyclable material collections.

As for vehicle emissions, reread the sections on vehicle emissions in Block 2: *Managing air quality* and think how many more vehicles are needed to carry out a recycling round as opposed to the more 'straightforward' waste collection.

Comparing the environmental benefits due to recycling with the disbenefits from additional vehicle (and processing) emissions has been extensively researched in recent years and will not be covered in depth here – suffice to say various life cycle assessments have been carried out in this area. The results are far from clear cut and the overall balance of benefits and disbenefits is strongly dependent on local circumstances.

3.3   Transfer stations

Waste transfer stations (WTS) are exactly that – points where waste is taken to be bulked up for transfer to landfill. RCVs are relatively small in terms of the loads they can carry – typically 9 to 12 tonnes, which represents the waste collected from some 600–800 houses. Disposal facilities are sometimes quite remote from where waste is collected, and to send every RCV to a remote site would be expensive in terms of staff, resources and fuel use.

A WTS will typically be located within or very close to a collection area. RCV staff will collect the waste, and then drive to the transfer station to discharge the load. As this is geographically much closer to the collection area than the disposal facility might otherwise be, they can get back to collecting much quicker, saving time, mileage and wear and tear on vehicles. The presence of a WTS means that an RCV and crew can typically serve 1200 households per day plus some commercial waste collections.

At the WTS, waste is bulked and further compacted, and loaded into large bulkers which can hold a mass of approximately 24 tonnes. Fewer trips are then needed to the disposal facility.

3.3.1   Health and safety issues concerning transfer stations

Mozzon et al. (1987) investigated the air quality in one transfer station during a summer period in the USA, concentrating on inorganic compounds such as respirable quartz, total dust and airborne metals. Bulk samples of soil cover and transfer station baghouse fines were also analysed for quartz, asbestos and PCBs. Asbestos and PCBs were not found in any samples. Quartz content of bulk samples varied from 8% to 31%. National Institute of Occupational Safety and Health (the US equivalent to our HSE) criteria of 0.05 mg m^{-3} for quartz was exceeded at two out of three transfer stations (particularly in the cabs of the WTS vehicles – loading shovels, etc.). It was concluded that significant exposures of total dust and respirable quartz were associated with these sites during warmer, drier weather, and further study of medical and hygiene aspects was recommended under a range of seasonal conditions in weather and waste composition.

You can experience an example of this on one of the Resources DVD video clips.

Higgins et al. (1987), as part of an investigation by the Health and Safety Executive in the UK, examined bioaerosols in seven transfer stations (additionally two incinerators, five landfill sites and a recycling plant for making refuse derived fuel, closed in spring 1986). Fungi and various types of bacteria were measured. At each site maximum levels of bacteria were 10^6, fungi 10^7 and actinomycetes (a type of thermophilic bacteria) 10^5 cfu m^{-3}. Opportunistic pathogens, such as *Klebsiella, Enterobacter* and *Aspergillus fumigatus* were found, but it was thought only susceptible individuals would be seriously affected whilst others would exhibit gastrointestinal symptoms. Concentrations were higher in summer months. However, *Penicillium sp.* and *A. fumigatus* were thought to be in concentrations that might penetrate to the alveolar spaces of the lungs and cause conditions such as asthma, and an infection in the lungs from the pathogen *A. fumigatus* (bronchopulmonary aspergillosis) (Higgins et al., 1987). Levels of Gram-negative bacteria and presence of pathogenic micro-organisms caused concern in the WTS.

Significant risks are associated with traditional waste handling facilities such as transfer stations. Physical hazards and accidents are thought prevalent (Cimino, 1975). Levels of heavy metals such as lead, chromium and mercury are reported in the literature. Bioaerosols are seen as a significant hazard, with levels as high as 10^7 cfu m^{-3}, although dust levels are also excessively high. Transfer stations and incinerators, in enclosed buildings, tend to have higher levels of bioaerosols, and the wearing of masks or improvement of hygiene levels is recommended in all of the studies shown.

3.3.2   Environmental issues concerning transfer stations

Transfer stations potentially have the same issues associated with them as many other waste management operations.

SAQ 8

What types of environmental issues might be associated with transfer stations?

Transfer stations are more likely to be surrounded by residential areas than most other types of waste management facility – as by their very nature they are placed in convenient locations for the bulking up of waste. However, very little research has been carried out to identify if this presents environmental issues to the surrounding population.

Crook et al. (1987) measured micro-organisms downwind of transfer stations, and found that although there were increases in numbers downwind compared with upwind (7.4 times more bacteria, 5.6 times more fungi), concentrations were only 10% of those found in the tipping halls within 50 m of the source. They concluded that the large concentrations of micro-organisms found in tipping halls rapidly dispersed when released into the atmosphere and that they were unlikely to cause problems for residents downwind of the sites. The potential dispersal of dusts and micro-organisms will be revisited later in much more detail in the section on composting.

4   HEALTH AND ENVIRONMENTAL IMPACTS OF WASTE RECYCLING AND COMPOSTING

4.1   Recycling

4.1.1   Processing collected materials

Once recyclable materials have been collected from the business premises, household or bring site, further processing is necessary before the materials can be transported to the reprocessor (glassworks, paper mill, etc.). Where the wastes are collected in a segregated form or sorted by the collection crew into individual waste types, all that is required is to bulk the material into loads that can be transported at a reasonable cost. At the other extreme, where mixed recyclable wastes are collected, there is a need for complex separation equipment. In either case this processing is carried out at a plant known as a materials recovery facility or MRF (often referred to as a 'murf'). MRFs are also referred to as 'materials reclamation facilities' and 'materials recycling facilities' in the literature and by the industry. Remind yourself about the configuration of simple and complex MRFs which are outlined in Section 11.5 of T210 Block 4: *Wastes management.*

The former Institute of Wastes Management (now the Chartered Institution of Wastes Management, CIWM) defines MRFs as:

> [A] central operation where segregated dry recyclable materials are processed mechanically or manually and the separation of recyclables prepares them to meet market specifications for sale.

(IWM, 2000)

The UK's waste strategies all call for a large increase in recycling, requiring additional MRFs. The English Waste Strategy (DETR, 2000) estimates that between 100 to 300 new MRFs with an average capacity of 40 000 tonnes a year will be required as a result. To a lesser extent, the continued increases in landfill taxes and the establishment of the Waste Resources Action Programme discussed below will also further increase tonnages entering MRFs.

As a result of these targets, the number of MRFs in the UK is already increasing rapidly (IWM, 2000) starting from only half a dozen in the late 1980s and early 1990s. At present virtually all local authorities either have an MRF or access to an MRF. The average number of employees in each MRF is between 11 and 19 people with the largest facilities having about 50 staff (IWM, 2000; Gladding, 2004). Therefore, there are in the region of 1100 to 1900 workers hand sorting waste in England and Wales at present, and this number is set to rise to 4400–7600 full-time workers. However, these figures represent maximum values; not all MRF operators are involved in hand sorting and the need for hand sorting in general may reduce as the technology advances.

There are four types of MRF, categorised on the basis of the types of waste they process:

- source-segregated household waste
- source-segregated household and commercial wastes
- commercial and industrial waste
- construction and demolition (C&D) waste.

MRFs that process household waste deal with what are known as 'dry recyclables', as outlined in the quote above. Dry recyclables are generally considered to consist of paper (newspaper, magazines), plastics (mainly PET, PVC, HDPE, occasionally polystyrene tubs), metals (food and drinks cans), textiles and glass (although fewer MRFs process glass due to the hazards associated with breakage). If the MRF also deals with commercial wastes this tends to be office and computer paper that can be sorted on the same paper lines, and cardboard which tends to enter via the tipping floor for immediate baling with little sorting required. Those that deal with just commercial and industrial waste tend to deal mainly in paper-based recyclables. Construction and demolition MRFs sort materials from building projects, i.e. concrete, metals, bricks, aggregate, rubble and wood, etc. MRFs could potentially process many different materials, or be single-stream and only process one type of material (such as a paper-sorting facility).

Materials recovery facilities are more often situated on industrial estates than some alternative waste management options as they are relatively odour free. Many are installed in existing warehouses, but many purpose-built facilities also exist. In some cases a MRF may also be attached to the side or on the same site as a waste transfer station, adjacent to a civic amenity site or on the same site as a landfill. In part this will depend on who owns and who operates the MRF – many different permutations are possible. It could be owned by a local authority, operated by a contractor or charity, or a partnership/joint venture might be in operation. This depends on the specifications of the original contract.

As shown in T210, a simple MRF may be no more than a collection of storage bays, conveyors and baling equipment, but the more complex MRFs will employ a range of technologies to sort materials including:

- eddy-current separators – to separate aluminium
- air classifiers – to blow lighter plastics from the stream
- inclined sorting tables – inclined conveyor with a curtain to separate the lighter fraction
- vibrating or oscillating screens – to separate larger and smaller items
- can flatteners/densifiers – to improve payloads for cans
- light separation equipment – uses infra-red light to identify and separate different types of plastics
- plastic bottle perforators – rotating drum with spikes to pierce bottles in multiple locations to prevent air taking up space when baled
- granulators – to chop plastic bottles into small chips.

All MRFs, however simple or complex, will generally use a system of conveyors, a magnetic separator and baling equipment, and will require people to either sort materials by hand or operate sorting machinery. The implications of this will be discussed later.

Some research and demonstration work was done on recovering materials from unsegregated municipal waste under the refuse derived fuel programme mentioned in Section 7. More recently, there has been some interest in mixed waste processing facilities (MWPF) (sometimes known as 'dirty MRFS'), but MWPFs clearly do not meet the CIWM's definition of an MRF quoted above. Also, the quality of the reclaimed products is generally lower, markets for the materials are limited and the impact on operators' health is potentially greater. However, mechanical biological treatment (MBT) plants that are also discussed in Section 7 do provide some opportunities for recovering materials from mixed wastes.

4.1.2 Markets for MRF outputs

Ever since the drive for increased recycling began in the early 1990s, the problem with what to do with the recovered materials has been present. Volatile markets for many materials has meant that it is difficult to plan long term for a recycling project. A typical example is cardboard where prices have varied widely over the past few years.

EXERCISE 2

Take a moment to look at current cardboard prices on the internet. The ROUTES page on the 'Library resources' link from the T308 website will give you a suitable website reference. In August and September 2004 prices were £45–48 per tonne. How different is the price now? In the past few years cardboard has varied from over £100 per tonne to less than zero (i.e. some dealers have charged suppliers to take cardboard away).

In order to stimulate the collection of recyclable materials, set standards for secondary materials and widen the markets for secondary materials, DEFRA established the Waste Resources Action Programme (WRAP). This way of directly addressing the problem of market barriers is in contrast to some other European countries which rely on state subsidies and regulatory approaches. WRAP identified four issues with markets for recyclables:

- *Fit for purpose.* Recycled materials are still widely perceived as second rate and the use of recyclate is often precluded by material-based rather than performance-based standards. Hence a 'standard' for materials is potentially needed.

- *Required supply.* There needs to be a demand for a material before it is produced, and preferably one that is of high enough value that it is worth collecting the materials. A consistent supply is also needed – this can be difficult in collection schemes which are dependent on what householders consume.

- *Demand.* A constant demand for materials is also needed. Once a recycling collection is established it is not possible to just turn off or reduce the supply.

- *Price/quality and value of materials.* Often sophisticated reprocessing technology is needed to deliver higher volumes, and quality often means that recycled materials struggle to compete with virgin materials.

In reality, recyclers are facing the same supply and demand issues any manufacturer faces – but with the additional problems that the feedstock is both less predictable and harder to control than when using virgin materials.

4.1.3 Environmental and health and safety considerations

In view of the expanding workforce employed in MRFs, studies investigating potential health effects on MRF workers are important. Household waste contains many different materials and therefore potentially numerous hazards, with sheer volume of waste exacerbating any difficulties. In recent decades, waste management has become progressively more mechanised with measures such as wheeled bins and mechanised lifters helping to separate and protect householders, collection crews and disposal site operators from the waste. However, the growth of recycling has tended to reverse this trend of distancing workers from the waste materials by bringing waste into closer contact with operatives during the hand sorting of materials at the MRF or during kerbside sorting.

SAQ 9

What hazards could result from the operation of an MRF? Think in terms of physical, chemical and biological issues.

Manual handling of materials and the ergonomic aspects of hand sorting are the main physical hazards, followed by the potential for accidents, e.g. cuts (broken bottles), broken limbs, etc., especially during interaction with heavy machinery and movement of vehicles. The Occupational Safety and Health Administration (OSHA) has implemented changes in the waste processing industry in the USA because of the serious issue of work-related musculoskeletal disorders (Krause, 2000). Further research on ergonomic issues had previously been recommended in MRFs (IEERR, 1995). Many MRFs are also vulnerable to potential fires. Noise and vibration are present in MRFs from vehicles, conveyors and baling machinery, and electromagnetic fields (EMFs) are potentially important due to the use of ferrous and non-ferrous separation equipment. In 1993 when MRFs were first introduced in the USA a 'health and safety manual' was produced encompassing such areas as manual handling, traffic, noise and fire (Fredrickson, 1992). A 1993 US Environmental Protection Agency publication provided a general overview of concerns when operating an MRF, from both an environmental and occupational perspective. It concluded that the environmental impacts of an MRF were likely to be very low (from fire, particulates, traffic, etc.).

Chemical hazards at MRFs include vapours and residues from household hazardous waste (HHW), e.g. garden chemicals, wood preservatives, paints, cleaning materials, etc. Heavy metals are included in this category due to the possibility of exposure to cadmium and mercury from batteries in HHW, an area first investigated in MRFs in Denmark, which showed some presence of mercury and lead (Sigsgaard et al., 1996). Volatile organic compounds (VOCs) are present in HHW and are also produced when waste is degrading (organic sulphur compounds for example), and are thought to contribute to complaints of nausea, irritation and intestinal problems experienced by some operatives (Wilkins, 1997).

Biological hazards have caused most concern in MRFs. As we saw earlier in Section 3.2, the collection and separation of household waste releases micro-organisms. This is also true of MRF operations where the dusts generated can contain airborne bacteria and fungi (bioaerosols), their cell wall components and airborne viruses. The microbial cell wall components are a particularly important constituent in organic dusts (Rylander et al., 1992). Among the most researched of these are bacterial endotoxins (a cell wall component in bacteria). Various respiratory symptoms and effects on the immune system have been seen from both endotoxin and glucan (Rylander, 1997a,b; Rylander, 1999). When micro-organisms are aerosolised, their viability decreases. Endotoxin and glucan have become more of a concern because they are released when a micro-organism dies. They are implicated in fever, flu-like symptoms, headaches, excessive tiredness and joint pains (termed 'organic dust toxic syndrome') and gastrointestinal problems (Rylander, 1997a,b). These symptoms have been reported in studies on waste sorting facilities (Poulsen et al., 1995a).

Additionally, it is important not to consider exposures as a single event. For instance, interaction between airborne exposures (bioaerosols and diesel particles and/or parameters such as VOCs) is a subject for future research in waste facilities, particularly focusing on the potential for inflammatory responses often seen in this occupation. Indeed, most volatile chemicals are capable of eliciting upper respiratory tract irritation (Dalton, 2002). Working outdoors in close proximity to vehicles may expose workers to diesel exhaust particles (DEP), or DEP may be drawn into vehicle cabs through open windows, doors or inefficient cab filters. DEP is known to cause irritation of the upper respiratory tract (Scheepers and Bos, 1992) and it has been recognised that this may contribute to health risks for waste handlers, both on its own and together with bioaerosols (Poulsen et al., 1995b).

4.1.4 Specific research studies

This section will give you an indication of a number of recent research studies on the environmental and health impacts of MRFs. This is a rapidly changing field, but the following paragraphs will help you to identify the key hazards, the risks they present and the issues relating to the need for standards.

In the USA a study on the health, safety and environmental issues associated with MRFs was initiated involving six sites covering a range of manual and mechanical waste segregation techniques (IEERR, 1995). Occupational measurements were made for dust, bioaerosols, heavy metals and noise. The study concluded that concentrations of heavy metals were generally very low or undetectable, and that concentrations of other potentially toxic elements such as arsenic, aluminium and nickel were also low. However, mercury concentrations reached 0.005 mg m^{-3} (sorting steel cans and aluminium) in some personal measurements (although these were more often undetectable). It was concluded that measurements of heavy metals in an occupational context would need further evaluation. Total suspended particulates were also measured at the six MRFs. Occupationally, personal total dust reached 2.5 mg m^{-3} but was often below 1 mg m^{-3}; respirable dust reached 0.57 mg m^{-3} but was more often below this.

Air sampling was carried out for viable organisms. However, despite the presence of some pathogenic species, the investigators considered the concentrations to be very low and unlikely to pose a significant risk to human health (typically of the order of 10^3 cfu m^{-3}). The USA study (IEERR, 1995) concluded that MRFs do not appear to pose a significant threat to public health or the environment, although it acknowledged that awareness of bioaerosols was increasing and that this area might warrant additional evaluation. It also acknowledged that there were no occupational exposure limits for bioaerosols, making it difficult to draw conclusions about effects on the health of the MRF operatives.

The same study undertook various environmental measurements. Concentrations of lead and mercury were very low. For particulates the environmental concentrations ranged up to 123 µg m^{-3} upwind (one measurement of 301 µg m^{-3}) and 139 µg m^{-3} downwind. PM$_{10}$ reached 107 µg m^{-3} upwind and 335 µg m^{-3} downwind. The study concluded that fugitive dusts might cause nuisance (although this could be mitigated through maintenance of roadways) and that heavy metals from an MRF made no significant impact on the surrounding community. In terms of bioaerosols, environmental concentrations did not differ significantly between upwind and downwind measurements: this suggests that the MRF added little to the pre-existing environmental impact of bioaerosols on surrounding communities.

A Canadian study investigated bioaerosols, gases, noise, EMFs and ergonomic factors at MRFs (Lavoie and Guertin, 2001). Many 'pollutants' were at levels not much above ambient, but the investigators found bioaerosols above their guideline of 10^4 cfu m^{-3} and concluded that workers' health could be at risk from these bioaerosol exposures. However, the main concern was ergonomics: workers were particularly liable to back pain due to standing and sorting activities.

In the UK, little work had been carried out on waste handling and health effects prior to 1996. Gladding and Coggins (1997) reported bioaerosol concentrations for individuals sorting recyclables in MRFs of up to 10^5 cfu m^{-3}, and symptoms reported by workers showed that 51% reported nasal irritation, 38% throat irritation, 21% eye irritation, 38% dry cough, 31% joint pains and 38% unusual tiredness. More recently Gladding (2002) and Gladding et al. (2003) reported that workers exposed to higher amounts of endotoxin and glucan when working in MRFs had an increased risk for respiratory symptoms compared with those with lower exposure – effectively a dose-response relationship was demonstrated. The interesting thing about this research is that workers who had been in the industry longer experienced the most symptoms (particularly gastrointestinal and respiratory issues), and that workers at MRFs fed by the twin wheeled-bin collection programme had a much poorer air quality (endotoxin and glucan exposures in excess of 10 ng m^{-3}) and so higher numbers of symptoms.

Many investigators including Sigsgaard et al. (1994a) tend to think that because of the relatively short existence of this industry, chronic health effects have not yet been reported. Symptoms most commonly seen in MRF research are pulmonary disorders, ODTS-like symptoms, gastrointestinal problems, eye inflammation, and irritation of the skin and upper airways. The term 'waste recycling worker syndrome' has been suggested for the fever, influenza-like symptoms, upper airway irritation and eye inflammation often seen in waste handling (Poulsen et al., 1995a). However, information on the hazards and their causes is still sparse. Poulsen et al. (1995a) concluded that levels of bioaerosols may reach 10^8 cfu m^{-3} during waste sorting and should be considered potentially harmful, and that plant sorting waste should be designed to prevent bioaerosol exposure. Specific activities such as manual sorting and baling, leading to greater aerosolisation, may pose higher risks for operatives. Indeed Germany previously introduced health regulations for all of its Duales System Deutschland plant for the protection of its 11 000 workers in sorting plant (TBRA, 1999). This specifies that workers should be separated from waste materials and that specific personal protection should be supplied where this is not possible. In conclusion, the health and safety of MRF workers is an important consideration when designing a recycling programme.

4.1.5 Research studies from mixed waste processing facilities

A substantial study of workers' health in mixed waste sorting facilities was carried out at a plant in Denmark in the late 1980s. The plant accepted household waste and industrial waste to a total of about 10 000 tonnes per year. The household waste was sorted by machine, the industrial waste partly by machine and partly by hand. Selected waste was converted into fuel pellets at the plant. The plant employed 20 people, 15 of whom were directly exposed to the materials to be sorted (in effect, mixed waste). Within months of opening, the number and type of the ailments among the operatives led the Danish National Environmental Protection Agency to initiate a study of working environments within these plants.

Of 15 exposed operatives, 5 were asthmatic, but others were also exhibiting flu-like symptoms, eye and skin irritation, fatigue and occasional nausea (Malmros, 1988). Microbial decomposition activity within the waste and endotoxins were suspected as a cause of these effects. Inspections at the site indicated proliferation of dust and food waste among 'sortable' materials, and accumulated wet refuse was sometimes mixed with material for reuse. Occupational medical measurements in September 1986 showed dust concentrations of 8.1 mg m^{-3} and total 'germs' of up to 3×10^9 cfu m^{-3}. Rebuilding of the plant was therefore undertaken in 1987–88; it included enclosing the conveyor belts and installing a central suction cleaning device in place of a compressed air blower. Other changes included a new ventilation system and microbial control programmes.

Medical studies of operatives from the original plant showed that eight operatives became ill within seven to eight months of the plant starting up. Nine cases of apparently occupational disease among the original 15 exposed operatives were eventually reported (Sigsgaard, 1990a; Malmros, 1992). The first two operatives to become ill were involved in cleaning and hand sorting. All symptoms began with eye irritation and sore throats, followed by respiratory symptoms, including chest tightness. Eight of the nine operatives were subsequently diagnosed with bronchial asthma; and seven of the nine operatives were accepted as having occupational diseases (Malmros, 1992). Infection was ruled out, as leucocyte counts (number of white blood cells) were normal. Eventually, seven of the nine operatives changed jobs, but only two out of the seven were free of symptoms after two years away from the plant (Sigsgaard, 1990; Malmros, 1992). This plant remains the most detailed investigation of the health of workers sorting mixed waste. As a result of this research mixed waste sorting was banned in Denmark.

In the UK, the sorting of mixed waste for materials recovery has not been actively taken up to any extent. This is partly due to health concerns, but also because of the questions over the quality of the outputs from these processes and the difficulty in finding markets for them.

4.2   Waste composting

It would be impossible for the UK to meet the requirements of its municipal waste management strategies without targeting the biodegradable (compostable) fraction of the waste. Diverting biodegradable waste from landfill by composting will also play a significant role in achieving the Landfill Directive targets. As a result, composting of biodegradable organic materials from our waste is becoming more common in the UK. Like all recycling operations, the aim is to generate a final product (in this case compost) which is no longer waste and has a beneficial use.

SAQ 10

List what we mean when we talk about compostable materials in our waste (you can refer back to T210 for information).

Making compost at home reduces the amount of waste for collection and treatment by the local authority. Therefore this operation is classed as 'waste reduction' in the wastes management hierarchy. If carried out correctly, composting allows the organic matter to decompose naturally in conditions of moisture and warmth through the action of micro-organisms. With sufficient aeration to maintain aerobic conditions and a well-balanced mixture of materials, this produces a soil-like substance rich in organic matter and containing plant nutrients. The decomposition process emits mainly heat and carbon dioxide. The process can take place in a simple heap, or more efficiently in a ventilated container. Both will build up sufficient heat inside the composting mass to kill or inactivate plant and human pathogens and weed seeds.

However, home composting relies on effort and commitment by the householders so participation rates tend to be low. If composting is to make a significant contribution to the recycling and Landfill Directive targets, compostable waste needs to be collected from individual premises (civic amenity sites, municipal parks and households) for central processing. This forms the subject of the remainder of this section.

If you refer back to Section 11.6 of T210 Block 4: *Wastes management*, the different methods of centralised composting are outlined. Remind yourself of the following methods:

You may also find it helpful to take a look at the composting video clip in the project material on the T308 Resources DVD.

- windrow
- static aerated pile
- enclosed composting (comprising containers, tunnels, agitated bays, rotating drums, silo or tower systems and enclosed halls).

It is important to note that composting sites can be owned and operated by a variety of organisations: private companies, local authorities, community groups and charitable concerns. Some sites also process a range of biodegradable wastes collected from households, industrial and commercial concerns, and municipal parks and open spaces. The process used often depends on local circumstances and the specification of the contract – for instance, a local authority might request a particular processing system or a site might only be able to support a particular system.

For the reasons I discussed with respect to recycling, the UK has experienced a massive growth in waste composting over the last few years. According to a survey carried out by the Composting Association the amount of material being composted in the UK increased 100% between 1999–2000 and 2001–02, with some 1.66 million tonnes of material processed at 218 composting facilities (Davies, 2003). In a recent government survey some 45% of local authorities in England and Wales reported that they were currently carrying out centralised composting, and a further 10% stated they were going to do so in the future (DEFRA, 2004a). In 1996–97, 84% of household waste was landfilled, in 2005–06 this had reduced to 62% and recycling/composting had increased from 7% to 27%. These figures do not include home composting, in which there has also been a substantial increase.

SAQ 11

Write down three key environmental benefits of making and using composts from wastes (if necessary reread Chapter 11 of T210 Block 4).

Despite the undoubted environmental benefits of composting there are a number of important environmental and public health implications:

- health and safety (collection of the material, and shredding/moving/screening it on site)
- downwind issues (particulates, dust and bioaerosols and other potential nuisance issues)
- 'normal' waste problems, e.g. vermin, noise, odour, traffic
- issues with the content of the compost itself – quality standards and the recent banning of catering wastes which may come into contact with meat
- environmental impacts of greenhouse gas emissions, leachate and nitrogen emissions.

We will now consider these in turn.

4.2.1   Environmental health and safety

For domestic waste, many composting schemes concentrate on green (garden and similar) wastes only and the industry as a whole is dominated by green waste composting, which accounts for approximately 80% of the raw feedstocks (by mass). A further 6% is food wastes, 3% kitchen and garden wastes and 11% a mixture of other organic materials including forestry, sewage sludge and paper/cardboard (Davies, 2003).

Where collection schemes exist, collection of the materials from the householders is dealt with by different local authorities in different ways. Some use a twin-wheeled bin system, with one bin being for compostable waste and the other for the remainder of the waste the household produces. Some areas supply a specific sack for householders to put their compostable waste in. There are various issues associated with collection of green wastes – if you need to, refer back to Section 3.2 to remind yourself of these.

Once the waste is collected it is delivered to a centralised composting site. There are, as a result of the waste, a number of potential hazards on a composting site with which an operator has to contend. If an HSE inspector were to visit the site, it is likely their main concerns would centre around traffic, manual handling and potential accident issues. However, there are many issues to consider on such a site.

SAQ 12

What health and safety issues do you think you would find on a composting site?

On any given compost site there is the potential for manual handling issues, particularly pushing and pulling bins of material (as outlined in Section 3.2 on collection). Transport issues arise from the use of machinery to turn and move material. Accidents can originate from foreign objects in material, e.g. syringes or broken glass, and many compost sites in the UK have suffered fires (mainly from spontaneous heating of material which smoulders, and when turned and exposed to air catches light). In a research report commissioned by the Environment Agency into compost sites, Wheeler et al. (2001) mentioned noise and vibration as additional potential hazards.

One of the main concerns on such sites is the presence of foreign objects and unwanted materials within the wastes collected for composting and in the product itself. This can include anything from batteries to household chemicals; public education campaigns and careful quality control procedures at civic amenity sites that collect compostable wastes are essential steps necessary to limit this problem. This also affects the quality of the compost (this issue is discussed below).

Volatile organic compounds (VOCs) and odours can originate from rotting plant materials and also the exhaust emissions from delivery vehicles and site engines (shredders, turners, screens, etc.) (Wheeler et al., 2001). Many VOCs carry occupational exposure limits and although Wheeler's study did not find significant concentrations on site, VOCs are often linked to odours, particularly of sulphurous compounds. These can cause workers to feel sick and nauseous. However, Swan et al. (2003) determined from the limited data available that VOCs were unlikely to be of major concern at compost sites, and this is supported in other waste management activities (Gladding et al., 2003).

Dust and micro-organisms such as fungi, bacteria and associated products are the focus of occupational health investigations at compost sites. Remind yourself about bioaerosols and exposure routes, which were covered in Section 3.2 on collection.

For each route of exposure a number of factors will need to be considered: whether effects are likely to be acute (short term) or chronic (long term); whether there is a risk of sensitisation (of respiratory tract if inhaled, or skin if contact occurs) or allergic reaction; whether it could be harmful to the reproductive process; or whether infection could be caused (such as in the case of micro-organisms). If we take recent HSE guidance relating to waste management (Swan et al., 2003) exposure to micro-organisms could cause ill-health in workers by inhalation and ingestion, via infection, allergy or an adverse response to toxins at compost sites.

Various studies have investigated this issue further. Ivens et al. (1997) found an association between stomach problems and exposure to bioaerosols at composting sites. Douwes et al. (2000) surveyed a small number of workers and discovered inflammatory changes in their upper airways. Bünger et al. (2000) found compost workers had more airways and skin symptoms than a control group of other workers. The study also found immunological changes and concluded that workers probably developed more effective immune systems as a result of their exposure. Wouters et al. (2003) assessed health symptoms by questionnaire at 31 compost sites in the Netherlands, discovering that respiratory symptoms were more prevalent in compost workers than in the general population. These workers also reported irritation of skin, nose, eyes and throat.

Swan et al. (2003) also highlighted the potential health problems from rats on such sites, e.g. *Leptospira*, the causative agent of Weil's disease. This bacterium multiplies in the kidneys of infected rats and is spread in contaminated urine, causing infection in humans through entry via skin abrasions and mucus membranes. Obviously if operatives are handling materials which also have the potential to break the skin then rats on such sites are of particular concern.

The health and safety of compost workers is of sufficient concern that it has also been addressed by the Composting Association with a specialist publication (Gilbert and Gladding, 2004). This covers various aspects of working at such a site, including: bioaerosols (it recommends workers are not located within 30 m of shredding, screening or turning); accidents; machinery; guarding; vehicles; electricity; noise; manual handling and the management of the facility, including relevant legislation and safe working practices such as active management of the material and safe use of personal protective equipment.

4.2.2   Downwind particulates and bioaerosols

Composting is classified as a waste recovery operation under the Waste Framework Directive. In practice, this means that it is carried out under a waste management licence issued by the national regulator or a licensing exemption registered with the regulator (for small facilities). The objective of both types of control is to ensure that the composting of waste is carried out in a way which protects the environment and human health.

The objective of the waste management licensing system is to ensure that waste management facilities do not cause pollution of the environment, do not cause harm to human health and do not become seriously detrimental to the amenities of the locality. More information on licensing can be found in the Legislation Supplement.

To this end the potential downwind exposure to dust and bioaerosols from waste composting facilities has become very important. Current Environment Agency guidance in England and Wales is that sites should avoid being located within 250 m of a sensitive receptor (Wheeler et al., 2001). Gilbert et al. (1999) developed a standardised protocol for monitoring bioaerosols downwind of composting sites based upon the following statement in section 5.93 of Waste Management Paper 4 which specifies:

> Turning of windrows leads to the emission of ... aerosols that may contain pathogenic micro-organisms. Such emissions are particularly difficult to control. Hence, unless the site is distant from sensitive receptors, the licensee should:
>
> - undertake background sampling for some time before operations begin;
> - monitor for airborne micro-organisms around the site.
>
> (DoE, 1988)

SAQ 13

What do you think a 'sensitive receptor' could comprise?

Identification of potential sensitive receptors is important. Various Town and Country Planning Acts (TCPAs) define these as:

> [A]ny building, other structure or installation, in which at least one person normally lives or works, other than a building, structure or installation within the same ownership or control as the operator.

A number of studies has been undertaken to determine whether micro-organisms travel downwind of composting plants. Epstein (1994) reviewed studies concerning sludge composting and concluded micro-organisms returned to background concentrations at approximately 1000 to 1500 m downwind of the site, much higher than other studies. However, Epstein (1994) concluded that the predominant amount of data from sewage sludge and solid waste composting facilities indicated that the dispersion of bioaerosols is primarily within the confines of the facility, at approximately 153 m. This is supported by Anon (1996) who found no impact on the residents of a community surrounding a compost site. Lavoie and Alie (1997) found that the quality of air 100 m downwind did not seem affected by recycling/composting operations.

Déportes et al. (1995) reported that aerial contamination may be ignored due to the relatively low exposure, and that compost dispersed in the air was mainly due to manipulation and so was an occupational problem only. This paper considered that application of composts by individuals to foodstuffs or application in public fields might pose a risk to health, with the most prominent risks being from hand-to-mouth contact by children. As there are crops in the vicinity of many composting operations this aspect should be considered; however, it should be remembered that crops are mainly subject to washing before consumption.

Danneburg et al. (1997) examined emissions of bioaerosols from an enclosed (in-vessel) composting plant where final screening took place out of doors. During screening, airborne bacteria concentrations reached 7.6×10^4 cfu m^{-3} close to the screen, reducing to 2.8×10^3 cfu m^{-3} 150 m downwind of the screen (the upwind concentration was 4.3×10^2 cfu m^{-3}). Concentrations of *Aspergillus fumigatus* (a pathogen often identified at compost sites) were reported at 2.0×10^3 cfu m^{-3} next to the sieve and 2.0×10^2 cfu m^{-3} 150 m downwind (none were found upwind). At a control location 3.1×10^2 cfu m^{-3} of bacteria and 78 cfu m^{-3} of *A. fumigatus* were found. Bioaerosols were detected in relation to the activities on-site which fed the in-vessel systems.

Schilling et al. (1999) reported that concentrations of fungi and *A. fumigatus* from an enclosed composting site were returned to background within 200 m, but that an open site showed levels in excess of background to 500 m. Neef et al. (1999) also reported excess concentrations to distances of 500 m.

Gilbert et al. (1999) investigated downwind measurements of bioaerosols at two composting facilities in the UK. In publishing a standardised protocol for measuring sites, it was determined that sensitive receptors were within 200 m of the boundary of the operational area of a composting site, unless complaints about emissions were located beyond this limit or local factors (e.g. local meteorological conditions) dictated otherwise. The 200 m figure was based on experience of dispersal monitoring during 1998. In an Environment Agency report, Wheeler et al. (2001) recommends a conservative guideline of 250 m, which forms the basis of current Environment Agency guidance (the distance at which a risk assessment of the site is required if sensitive receptors are in the vicinity). It is at these distances that Wheeler et al. (2001) determined that bioaerosols would be reduced to background, with guidance utilising a conservative concentration of 1000 cfu m^{-3} for bacteria and fungi, effectively a no-observed effect level (i.e. a concentration or dose of bioaerosols that has been reported to have no harmful (adverse) health effects), and these exposure guidelines are further outlined in the Environment Agency Guidance Note M17 (Environment Agency, 2003) on particulates emitted from waste facilities. These concentrations were reviewed and further agreed with by Swan et al. (2003) in a report for the Health and Safety Executive. However, these views are relatively conservative as they rely on straight line modelling of bioaerosols (when die-off leads to an exponential decrease) and conservative exposure values.

Hryhorczuk et al. (2001) found that on-site concentrations of total bacteria (7.9×10^4 cfu m^{-3}) demonstrated a statistically significant pattern of decreasing concentration with distance from piles of compost and demonstrated higher downwind versus upwind concentrations. As in previous studies, concentrations were higher during activity on-site. The most common species of fungi recorded were *Aspergillus, Penicillium and Cladosporium.* Masks were recommended for workers, along with wetting of the compost to reduce dust generated.

Epstein et al. (2001) found that the effective management of dust significantly reduced the release of the human pathogen *Aspergillus fumigatus* from a composting facility during the construction of windrows, turning and screening processes (Figure 4).

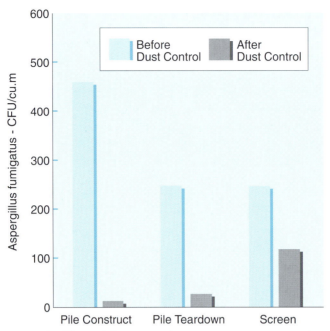

Figure 4   Effect of dust control at a composting site on the release of *Aspergillus fumigatus* (source: Epstein et al., 2001)

This is important since research has shown that when there is little activity on-site then little bioaerosol or dust release occurs and it is these activities that generate dust and initiate the concerns associated with bioaerosol release (Gilbert et al., 1999). Hence if a site uses dust suppression techniques to prevent nuisance dusts, their potential generation of dust and bioaerosols greatly decreases. Wheeler et al. (2001) also recommend that compost be kept moist to reduce production of dust and hence bioaerosols.

In terms of studies to see if bioaerosols affect the health of people living around the sites, there are very few – Browne et al. (2001) reported no increase in allergies and asthma symptoms in people living around compost sites, but Herr et al. (2003) showed a significantly elevated risk for bronchitis, colds, eye irritation and general health. However, as the data are currently so limited, it is not possible to draw definitive conclusions as to whether sites do present a risk to the surrounding population.

In 2004 DEFRA carried out a survey of studies to determine the health effects associated with a range of waste management activities, including composting. The study was a literature review incorporating many of the studies mentioned above, and concluded that composting was an area that needed further investigation, and that increased emissions under non-standard operating conditions could be of concern for open windrow composting (DEFRA, 2004b). Hence this issue is likely to be further investigated in the future.

In summary, the majority of studies carried out at composting sites were done at open-air facilities (windrow systems). Other types of compost system such as in-vessel plants are enclosed but can still generate bioaerosols downwind during operations such as loading and unloading the vessels, shredding and screening. The general consensus is that at sites dealing with compostable wastes concentrations of bioaerosols return to background within 200–250 m of the source. Although the presence of micro-organisms from a site may be reported at distances up to 1000 m downwind, the concentrations generally reported are low at these distances. From these studies it is clear that compost sites do generate bioaerosols in excess of background but they are dispersed relatively rapidly.

In terms of 'natural' concentrations of bioaerosols, Hryhorczuk et al. (2001) during an investigation of a windrow composting site, reported high measurements of fungi off-site in a wet woodland comparable to on-site. Additionally, it was reported that mowing a nearby meadow also significantly affected results of viable fungi and bacteria (160 cfu m^{-3} and 480 cfu m^{-3} respectively prior to mowing, 15.0×10^3 cfu m^{-3} and 17.6×10^3 cfu m^{-3} after). This study demonstrates how other activities and environments can generate bioaerosols in excess of background, and certainly in concentrations similar to composting. Indeed, agricultural activities such as farming (either grain or cattle) are reported to generate concentrations in the order of 10^5 cfu m^{-3} (harvesting 10^7–10^8 cfu m^{-3}) for both bacteria and fungi (Swan et al., 2003). Additionally, other authors report that natural concentrations of bacteria and fungi routinely range from 1000 to 100 000 (10^3 to 10^5) cfu m^{-3} air (Cox and Wathes, 1995). What we must not forget is that composting is a natural activity – and the bioaerosols generated are also natural. It is only when significant exposures are experienced that problems can occur. This leads us on to what are 'acceptable' levels of exposure to dusts and bioaerosols.

Hence for bioaerosols, the Environment Agency adopts a risk-based approach to allowable concentrations. The allowable value in any case will depend on existing background levels and (for perimeter fence monitoring) the distance to sensitive receptors. Limit values quoted in Table 12 are 'yardstick' values only.

Table 12 Recommended guidelines for environmental contaminants from compost sites

Application	Determinand	Limit value	Reference
Annoyance	Deposited dust	200 mg m^{-2} per day	Custom and practice
	Soiling index	25 soiling units (SU) per week	Custom and practice
Environmental protection	Suspended particulates	24-hour average PM$_{10}$ concentration not to exceed 50 μg m^{-3} more than 35 times per year.	EU limit and National Air Quality Strategy objective
Air quality management		Annual average PM$_{10}$ concentration not to exceed 40 μg m^{-3}	
Environmental protection	Bioaerosols – bacteria	8-hour average not to exceed 1000 colony forming units suspended per cubic metre air (cfu m^{-3})	Custom and practice
	Bioaerosols – Gram-negative bacteria	Not to exceed 300 cfu m^{-3}	Custom and practice
	Bioaerosols – fungi	Not to exceed 1000 cfu m^{-3}	Custom and practice

Source: Environment Agency (2003a).

There are no standards for concentrations of total suspended particulates (TSP), although a guideline was suggested for nuisance purposes and Wheeler et al. (2001) suggested concentration guidelines at 10% and 2.5% of the occupational exposure limit for inhalable dust (10 mg m^{-3}), i.e. 1000 µg m^{-3} and 250 µg m^{-3} respectively.

4.2.3   Other environmental impacts

Swan et al. (2003) illustrate the role odour has to play in downwind exposure to dust and bioaerosols from a composting site, which can be a nuisance to nearby sensitive receptors. Odours from composting occur when anaerobic microbial activity takes place causing the release of mainly sulphurous compounds, which can result in complaints from neighbours near to the site. Residents believe that as they can 'smell' a site then there are dust and bioaerosols present that can affect their health, whereas in fact odour potentially travels further and is detectable in quantities much lower than would have an effect on health. Hence odour is of potential nuisance value at composting sites, even if the concentrations are not in the range that might affect health. Indeed, Wheeler et al. (2001) state that 'odours are not directly a risk to health and can more correctly be described as a nuisance'.

4.2.4   Compost and composting standards

Composts have a variety of uses that may have a potential impact on the health of the public and on the environment. In particular, composts may be used in growing food crops and in domestic gardens where small children can come into contact with them. The following potential hazards and nuisances could arise from using composts.

- the accumulation of heavy metals and other potentially toxic elements in the soil

- the presence of human pathogens (such as *E. coli* and *Salmonella)*

- plant pathogens such as those responsible for potato blight

- viable weed seeds and plant material capable of growing (particularly those classed as 'noxious weeds' such as ragwort and Japanese knotweed)

- other physical contaminants, e.g. plastics, sharps, etc.

Therefore it is essential that composts are produced to standards which ensure that the product does not have any adverse effect on human health or the environment. A number of national and international groups have been working in this area for many years and several standards have been suggested.

In the UK, the Waste Resources Action Programme (WRAP) (See Section 4.1.2) identified the absence of a nationally recognised standard as a major barrier to developing markets for composts. Therefore WRAP funded the development of a standard, known as 'PAS 100' issued by BSI (British Standards Institution, 2002). Although not a British Standard itself, PAS 100 specifies process controls, record keeping, input materials, sanitisation, stabilisation and quality requirements. The emphasis is on identifying and managing the process; obviously this requires the operators of compost sites to keep detailed records. PAS 100 also requires product identification and traceability.

For full details of the requirements of PAS 100 you should consult the above reference. However, you should be aware of the following key points.

■ To comply with PAS 100, the compost must only be made from source-segregated wastes; composted mixed waste or material segregated from mixed waste by manual or mechanical means will not meet PAS 100.

■ The quality requirements listed in Table 13 must be met.

Table 13 PAS 100 quality requirements

Human pathogens:	
Salmonella spp	Absent in a 25 g sample
Escherichia coli	<1000 cfu g^{-1}
Potentially toxic elements:	
Cadmium	$\leqslant 1.5$ mg kg^{-1} dry matter
Chromium	$\leqslant 100$ mg kg^{-1} dry matter
Copper	$\leqslant 200$ mg kg^{-1} dry matter
Lead	$\leqslant 200$ mg kg^{-1} dry matter
Mercury	$\leqslant 1$ mg kg^{-1} dry matter
Nickel	$\leqslant 50$ mg kg^{-1} dry matter
Zinc	$\leqslant 400$ mg kg^{-1} dry matter
Physical contaminants:	
Total glass, metal and plastic >2 mm	$\leqslant 0.5\%$ of which $\leqslant 0.25\%$ is plastic (by mass air dried basis)
Stones and other consolidated mineral contaminants >2 mm	$\leqslant 7\%$ (by mass air dried basis)

Source: British Standards Institution (2002).

Following the foot and mouth disease outbreak in 2001 the composting of waste from kitchens and catering establishments was banned. This ban was introduced due to the risk that such wastes could be contaminated with meat products, and covered all domestic and commercial kitchen waste (including those arising in vegetarian kitchens). This blanket ban was relaxed in 2003 through the introduction of the Animal By-Products Regulations (HMSO, 2003). Under these regulations, only green (i.e. non-kitchen) waste may be composted in windrows. Material containing kitchen and catering waste may be composted, but only in enclosed two-stage systems which raise the composting material to a specified temperature for a given time (the values depending on the system used).

In addition, the regulations:

■ specify measures to be taken to eliminate the risk of contamination of the compost with the untreated feedstock

■ require sites to be approved by the State Veterinary Service before handling and processing the materials covered by the regulations

■ require hazard analysis and critical control point (HACCP) analysis, which requires monitoring and checking of the plant and procedures, and other requirements such as processing of newly received material within 24 hours

■ require provision for clean areas and for vehicles and containers to be cleansed and disinfected.

Home composting is not covered by these regulations so householders wishing to compost their own kitchen scraps on their own compost heap are exempt from the rules, provided that they do not keep pigs, ruminants or poultry on the premises. However, the disposal of meat scraps in garden compost heaps is not recommended in any event. Doing so can increase the risk of spreading disease via scavenging wildlife to livestock.

Separate guidelines from a risk assessment report were published for the composting of catering wastes (Gale, 2002). The project considers a wide range of human and animal pathogens, including:

- transmissible spongiform encephalopathies (TSEs) including BSE and scrapie

- exotic diseases such as foot and mouth disease, classical swine fever and Newcastle disease

- endemic diseases such as *E. coli* 0157, campylobacters and mycobacteria.

This report concluded that the risks to the food chain were very low, and that application of the resultant compost to vegetables grown for human consumption were also low. It concluded that composting could be safely done providing suitable safeguards and treatment standards were introduced. Indeed, the report outlined the fact that landfilling could in fact present more of a risk to surrounding animal herds. However, compost derived from catering waste may only be used on non-pasture land. Non-pasture land is land that has not been grazed for at least two months after application of the compost. It is anticipated these measures will reduce risk of disease to farm animals. The risk assessment also considered the risks to plant health in principle, but more detailed work remains to be done to assess the risk to crop production. Meanwhile, plant health risks are managed by following existing guidance on composting as published in DEFRA's Plant Health Code of Practice for the Management of Agricultural and Horticultural Waste (PB 3580).

There will, however, be guidelines for dealing with catering wastes. The EU standard is treatment at 70 °C for 1 hour with a maximum 12 mm particle size. However, allowing only one treatment standard would restrict the number of premises able to comply with the new rules. Therefore DEFRA suggests that the risk assessment has demonstrated that alternative methods to the EU standard can achieve equivalent pathogen removal. This option would allow plants to meet one of the national standards, or they may choose to comply with the EU standard. Allowing a range of treatment standards maximises the number of different composting and biogas systems used by industry that will be able to conform with the new rules. It also ensures that the UK is flexible enough to meet EU landfill targets.

4.2.5   Wider environmental impacts

As discussed above, the most noticeable environmental impacts of composting relate to local issues, but it is important to consider the wider impacts. As I stated in Section 2, the main potential impacts are gaseous emissions and leachate.

Gaseous emissions

Composting releases carbon dioxide; published values range from 100 kg per tonne of feedstock to 323 kg per tonne of feedstock (DEFRA, 2004b). Although CO_2 is a greenhouse gas, it should be noted that this emission is a re-release of CO_2 that was absorbed from the atmosphere during the growth of the plant material being composted. Therefore composting does not cause a net contribution to atmospheric CO_2 levels. Indeed it can be argued that if not composted, the material would be landfilled leading to the generation and release of methane, a much more potent greenhouse gas.

Hobson et al. (2004) investigated the emissions of the greenhouse gases methane and nitrous oxide (N_2O) from the surface of a windrow formed from green waste that had already been partially composted in an enclosed vessel system. The fluxes (emissions per unit area of windrow) are shown in Table 14.

Table 14   Emissions from composting windrows

Day	CH_4 flux (mg m^{-2} h^{-1})	N_2O flux (mg m^{-2} h^{-1})
7	6.604	0.370
14	4.142	0.271
21	1.055	0.006
35	6.120	0.029
50	5.020	0.627
64	0.862	0.030
78	0.050	0.005
92	0.215	0.025

Source: Hobson et al. (2004).

Hobson et al. concluded that the methane release was due to the formation of anaerobic pockets within the windrow. This was in spite of the regular turning of this experimental windrow. The low nitrous oxide releases were considered to be due to the fact that much of the volatile nitrogen in the feedstock would have been emitted in the form of ammonia.

DEFRA (2004b) cited ammonia emission rates of 5–120 g per tonne of feedstock, but added that the higher figures related to whole waste composting. This has the potential for local environmental impacts, particularly if the facility is close to a natural landscape with low nutrient soils. The deposition of ammonia in such areas may lead to nitrogen enrichment of the soil, which can disturb the local ecosystem by allowing plant species that can tolerate low nutrient soils to be displaced by more vigorous species that benefit from the higher nitrogen levels.

SAQ 14

If the mean methane flux from a windrow is 3 mg m^{-2} h^{-1}, how much methane would be released from a 100 m long windrow 2 m in height and 4 m wide having a triangular cross-section if composting continues for 90 days?

Leachate and run-off

Leachate consists of rainwater that has percolated through a compost windrow and water present in the composting waste that has drained from the base of a windrow. Typical concentrations of potential pollutants are given in Table 15 (Roth, 1991). Run-off is rainwater that has fallen onto a composting area but hasn't percolated through the windrows (CIWM, undated). For most outdoor windrow systems, leachate and run-off mix on the composting area and are collected as one effluent stream.

Table 15   Compost leachate composition

Parameter	Value
BOD_5	10 000–46 000 mg l^{-1}
COD	18 000–68 000 mg l^{-1}
NH_4^-	400–1100 mg l^{-1}
Total nitrogen	500–2085 mg l^{-1}
pH	5.7–10.2

Source: Roth (1991).

Clearly, leachate and run-off have a potential to pollute water courses, for example if this effluent was discharged into a local stream or river. However, provided that composting is carried out on a concrete pad and the leachate is collected it is unlikely to present a serious environmental threat. In fact, most composting operations are able to use the leachate collected during wet periods to maintain adequate moisture levels in the windrows during dry periods. Therefore, most sites release little if any leachate.

5   MUNICIPAL WASTE INCINERATION

> **READ**
>
> The design and operation of municipal waste incinerators is reviewed in T210. You should reread Chapter 9 of T210 Block 4: *Wastes management* at this point.

Incineration does not present a total solution to the waste management problem; to meet national recycling targets it must be combined with recycling and composting operations. The role of incineration in integrated waste management is discussed in Chapter 14 of T210 Block 4: *Wastes management* and developed further in Section 8 of this block.

In summary, incineration is a well-established technology. Plants can be designed and built with a guarantee that particular thermal output and environmental emission standards will be achieved. However, incineration is not without problems. A number of national and multinational pressure groups have adopted policies of opposing all incineration processes as a matter of principle. Consequently, proposals for new incinerators are always opposed.

Also, there is usually strong local opposition to the development of new incinerators. In common with opposition to landfills, recycling and composting facilities, local objections tend to be based on potential problems such as odours, traffic, visual impact and impact on property prices. Careful siting of facilities, their design and operation can help to minimise the impact of a waste management operation but, as with any industrial development, it would be unrealistic to expect no opposition.

EXERCISE 3

Read File Articles 1 and 2 of Appendix 1 ('The next big thing' and 'Burn me') to see two contrasting views of incineration. *Resource*, the publisher of 'The next big thing', describes itself as a 'not for profit magazine aiming to take waste management forward by providing new ideas and information on sustainable resource management'. The article was written by the magazine's editor. *New Scientist*, which published 'Burn me', is a weekly magazine of news and articles aimed at those in the scientific community. The author of the article is a journalist specialising in environmental issues who is not afraid of generating controversy. Now summarise the arguments against incineration from Article 1 and how these arguments are countered in Article 2. Note any inconsistencies or flaws in the arguments provided in either article.

Article 1, 'The next big thing', in Appendix 1 suggests that the UK could recycle much more waste than it currently does and that there would be economic benefits in doing so. It goes on to imply that up to 80% of our waste could be recycled, reducing the need for incineration to meet the requirements of the Landfill Directive; it adds that incineration is the main source of dioxin in the UK and that there is a direct conflict between recycling and incineration.

In contrast, Article 2, 'Burn me', states that recycling is only carried out because it makes people feel good and that the economic and environmental arguments for doing so are unsound. In particular, it cites a study that claims incineration to be 'better' than recycling when it comes to paper. 'Burn me' then goes on to note that much of the CO_2 generated from incineration is renewable (i.e. derived from plant material), so leads to a reduction in atmospheric CO_2 levels if the power generated from the process displaces that produced by burning fossil fuels.

So what can we deduce from these articles?

Firstly, you will recall from T210 Block 4: *Wastes management* that published values for the amount of household waste that can be recycled range from 46% to 70% while in SAQ 20 of that block we estimated that about 53% of household waste could theoretically be recycled. This suggests that 'The next big thing' is perhaps being rather optimistic in its suggested recycling rate of 80%. In Chapter 14 of T210 Block 4 we also demonstrated that recycling and incineration can work together – as long as both schemes are correctly sized. As you saw in Section 4.2.1 of T210 Block 4, the latest data on dioxins (DEFRA, 2004b) suggest that the whole of the wastes management industry accounts for 1% of the UK's dioxin emissions. ('The next big thing' probably took its information on emission levels from the older, now closed, incinerators that only had the most basic combustion controls or gas cleaning equipment.)

'Burn me' does tend to ignore that the national recycling targets are a fact and that it would be unrealistic to suggest that they will be abandoned. The article neglects to note that as well as the 'feel good factor', the action of recycling helps to increase the general awareness of waste and the wider issues of sustainability.

Both articles claim the environmental and economic high ground. In Section 1.2.7 I summarised the range of costs produced for DETR which show that the situation is much more confused than either article suggests. When external costs are also taken into account the picture is even worse, with the need to place monetary values on (say) a tonne of emitted CO_2 and the loss of a hectare of peat beds. So when it comes to costs all we can say is that there is a great deal of uncertainty and we need to treat all figures (especially those from sources with a particular axe to grind) with extreme caution.

5.1   Incineration emissions and their control

> **READ**
>
> Section 9.5 of T210 Block 4: *Wastes management* describes the main systems to control the emissions from MSW incinerators. If you are not familiar with these technologies, reread Section 9.5 before working through the material below.

5.1.1   Emission limits and current performance

The daily average limit values for atmospheric emissions as specified in the Waste Incineration (England and Wales) Regulations 2002, which implement the Incineration Directive of 2000, are shown in Table 16. This table also shows the values achieved by three installations:

SNCR involves injecting ammonia or urea into the furnace in the temperature zone where the ammonia (or urea) reduces the NO_x to nitrogen and water. This process is described in Section 9.5.4 of T210 Block 4: *Wastes management*.

■ the Chineham plant in Basingstoke. This facility was commissioned in 2003 and designed to meet the daily average limit standards specified in the table. The flue gases are treated in a semi-dry scrubber (described in Block 2: *Managing air quality*) followed by a bag filter and selective non-catalytic reduction (SNCR) using urea injection to control NO_x

■ the SELCHP plant in London which was commissioned in 1994 and was designed to meet the requirements of the 1996 incinerator directive. It has a similar gas cleaning system to Chineham, but was not originally fitted with NO_x control equipment (note that an SNCR system was installed at SELCHP in 1998 to achieve compliance with the current standards)

■ the Spittelau incinerator in Vienna fitted with wet scrubbing and selective catalytic reduction (SCR) to reduce NO_x emissions.

Table 16   MSW incineration emission standards and levels

Pollutant	Daily average limit	Units	Values achieved		
			Chineham	SELCHP	Spittelau
Total dust	10	$mg\ m^{-3}$	4	<3	0.05
Total organic carbon	10	$mg\ m^{-3}$	0–1	<3	<1
CO	50	$mg\ m^{-3}$		15	45
HCl	10	$mg\ m^{-3}$	7–9	<5	<0.8
HF	1	$mg\ m^{-3}$		<0.1	<0.02
SO_2	50	$mg\ m^{-3}$	7–15	20	7.9
$NO + NO_2$ expressed as NO_2	200	$mg\ m^{-3}$	163–191	320	17
Compounds of Cd and Tl	0.05	$mg\ m^{-3}$		<0.01	
Compounds of Hg	0.05	$mg\ m^{-3}$		<0.01	
Compounds of $Sb + As + Pb + Cr + Mn + Ni + V$	0.5	$mg\ m^{-3}$		<0.1	
Dioxins and furans	0.1	$ng\ m^{-3}$ (TCDD equivalent)		<0.05	0.03

Chineham figures represent the range of daily averages for April 2004.

SELCHP data refer to a period before the introduction of the limits shown.

Vienna plant fitted with wet scrubbers and SCR De-NO_x. Source: Rabl et al. (1998).

TCDD equivalent expresses the total dioxin and related compound concentration by equating it to the concentration of 2,3,7,8 tetrachlorodibenzodioxin that would have the same toxicity as the total emission.

The key points to note from this table are that while the wet scrubbing system clearly achieves a far higher reduction in emissions than the other two facilities, they all easily meet the acid gas, particulate material, metals and dioxin standards. The NO_x levels at SELCHP are typical of the values achieved with simple flue gas recirculation whilst Chineham's values typify the performance of SNCR and the much lower emissions at Spittelau demonstrate the effectiveness of SCR.

Specifying emission limits is an effective way of demonstrating that the plant is designed to meet BAT (best available techniques) and that the plant is operating correctly. However, in terms of environmental impact, it is important to consider the impact of the emissions on local air quality.

EXAMPLE I

For the installation described in the table below, use the 'point source' model given in the Air Quality Management Tools on the Resources DVD to estimate the maximum short-term concentration of NO_x due to the operation with emissions equivalent to those given above for no NO_x control, SNCR and SCR.

Parameter	Value
MSW throughput	10 t h^{-1}
Effective chimney height (taking account of the buoyancy of the stack gases)	100 m
Wind speed	5 m s^{-1}
Flue gas flow rate	16.67 Nm^3 s^{-1}
Atmospheric stability class	D (the most common in the UK)

ANSWER

We could carry out the calculation three times, using the three emission rates given, but it is easier to begin by assuming an emission concentration of 1 mg m^{-3}.

Therefore the mass emission (total gas flowrate × emission concentration) is 16.67 × 1 (i.e. 16.67 mg s^{-1}).

Using these data in the Gaussian Plume Model, which is provided on the DVD and uses the equation

$$C(x,y,z) = \frac{Q}{2\pi u \sigma_y \sigma_z} \exp\left(-\frac{y^2}{2\sigma_y^2}\right)\left(\exp\left(-\frac{(z-H)^2}{2\sigma_z^2}\right) + \exp\left(-\frac{(z-H)^2}{2\sigma_z^2}\right)\right)$$

gives the graph shown in Figure 5.

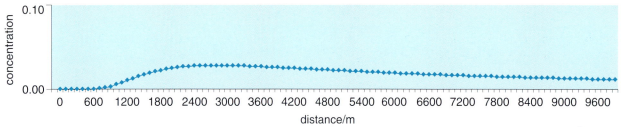

Figure 5   Pollutant dispersion: concentration under plume centre line for emission concentration of 1 mg m^{-3}

From the model output, we see that the maximum ground level concentration is 0.0028 μg m^{-3} and that this value occurs at about 2.9 km downwind of the source. However, this is for an emission concentration of 1 mg m^{-3} so for our three cases the maximum concentrations are:

for an emission of 17 mg m^{-3} (SCR): 0.0028 × 17 = 0.048 μg m^{-3}

for an emission of 191 mg m^{-3} (SNCR): 0.0028 × 191 = 0.53 μg m^{-3}

for an emission of 320 mg m^{-3} (no NO_x control): 0.0028 × 320 = 0.90 μg m^{-3}

Referring back to T210 Block 6: *Air quality management,* we see that the World Health Organization and the EU have set air quality targets for NO_x of 200 μg m^{-3} as an hourly average. In the worst case, the incinerator emissions account for around 5% of this limit whilst, for a plant with SNCR, the emissions account for about 3% of the limit.

SAQ 15

Use the same model to estimate the maximum ground level concentrations of sulphur dioxide and HCl due to the operation of a 10 t h^{-1} MSW incinerator operating according to the current emission limit values.

5.1.2 Quantities of air pollution control residues

The design of gas scrubbing equipment is highly specialised and involves the use of complex computer modelling techniques. The role of the generalist waste manager is to verify that the performance offered by a particular design is realistic and that the mass balances that the design is based on are sound.

SAQ 16

(a) A 10 t h^{-1} MSW incinerator produces 16.67 Nm^3 s^{-1} of combustion gas which is treated in a dry scrubber and bag filter. If the raw waste contains 0.8% chlorine and 90% of it enters the gas stream in the form of HCl, calculate the flue gas HCl concentration (in mg Nm^{-3}).

(b) Lime Ca(OH)$_2$ injection is used to reduce the HCl concentration to 10 mg m^{-3}. Calculate the amount of lime used if the scrubber design calls for the injection of twice the theoretical amount. (Hint: you will use the equation Ca(OH)$_2$ + 2HCl $\longrightarrow$ CaCl$_2$ + 2H$_2$O)

(c) What is the total mass of solid residue produced from this process?

(d) If the flue gas particulate material content is reduced from 7000 mg m^{-3} to 10 mg m^{-3} what is the mass of collected particulate material in the scrubber residues? (Atomic masses Cl = 35.5, H = 1, Ca = 40, O = 16)

5.1.3 Management of solid residues from incineration

Although incineration reduces the volume of waste for disposal by 90% and the mass by about 70%, this still leaves an appreciable amount of material for disposal or recovery. To develop an ash management strategy requires an understanding of the physical and chemical properties of the materials and the regulations governing their disposal. Two key questions that must be answered are:

■ Is the material a hazardous waste?

■ Will the material be acceptable for landfill disposal?

To answer these we need to consider the requirements of the Hazardous Waste Directive (Council of the European Communities, 1991) and the Waste Acceptance Criteria (Council of the European Union, 2003).

The Hazardous Waste Directive is intended to supply a method for determining whether any given waste is hazardous or not and to ensure that such wastes are managed correctly. In support of the first of these aims the directive states that a waste is to be classed as hazardous if it has one or more of a list of 14 hazardous properties (for example, explosive, oxidising, flammable, toxic, carcinogenic, etc.). Subsequent directives have helped to clarify the procedure for classifying wastes by means of the European Waste Catalogue (EWC) (Environment Agency, 2003b).

The latest version, EWC 2002, is a comprehensive list of all wastes categorised by the origin of the waste. All wastes on this list are classified as either:

- non-hazardous waste
- hazardous waste (absolute entry) – these wastes are classed as hazardous regardless of concentration of hazardous substances they contain;
- hazardous waste (mirror entry) – these wastes are classed as hazardous or not, depending on the concentration of hazardous substances present.

The Waste Acceptance Criteria (WAC; Council of the EU, 2003) set out the criteria for the deposition of waste at the three classes of landfill site (inert, non-hazardous and hazardous). For a material to be accepted at a hazardous waste site, the total organic carbon content must be less than 6% or the loss of ignition must be less than 10%. Additionally, the leachability of specified heavy metals and halogens must be below specified levels when determined in a standard leaching test. The WAC also specifies maximum leachabilities for wastes to be deposited at non-hazardous and inert waste landfills.

In the following sections I will review the properties of MSW incineration residues and their classification under these regulations.

5.1.4   Bottom ash composition and management

In 1994/95 WRc plc, working with the Building Research Establishment (BRE) and the Netherlands Energy Research Foundation (ECN), undertook a programme of incinerator residue sampling and characterisation for the Department of Trade and Industry's Energy Technology Support Unit (ETSU) (Lewin et al., 1995, 1996). The following paragraphs are based on the findings of this research programme and describe the properties of bottom ash and air pollution control residues obtained from a UK incinerator burning municipal solid waste.

Physical characteristics

The bottom ash was found to be heterogeneous, mainly consisting of ash, but with small stones, grit and glass particles, some unburned material and small metal objects. After crushing a sample, 31% by mass was still greater than 2 mm and consisted of bricks/stone, clinker, ceramics, glass and metal.

Separate samples were subjected to size distribution and bulk density measurements to determine whether the material complied with the requirements of BS 3797 for lightweight aggregates. Most samples complied with BS 3797 (British Standards Institution, 1990) and those that did not would have done after a limited amount of crushing and blending.

Chemical composition

Chemical and mineralogical analyses were undertaken and the results are summarised in Table 17.

Table 17   Composition of bottom ash

	Units	Value
pH		10.4–12.1
Moisture	%	15–17
Loss on ignition	% (dry)	3.1–5.2
Carbon	% (dry)	2.0–2.9
Si	% (dry)	19–22
Al	% (dry)	2.1–3.9
Mg	% (dry)	0.84–1.0
Fe	% (dry)	1.9–2.8
Ca	% (dry)	10.1–12.4
Cl	% (dry)	0.28–0.52
SO_4	% (dry)	2.2
F	% (dry)	0.03
Zn	ppm (dry)	2898–3790
Pb	ppm (dry)	2632–2850
Cu	ppm (dry)	1330–2510
Mn	ppm (dry)	811–876
Cr	ppm (dry)	260–307
Cd	ppm (dry)	<1–12
As	ppm (dry)	13.6–15.3
Mo	ppm (dry)	<10
Hg	ppm (dry)	<0.25
Main minerals identified		quartz (SiO_2), calcite ($CaCO_3$), anhydrite ($CaSO_4$), gypsum ($CaSO_4.2H_2O$)

Source: Data from Lewin et al. (1996).

Leaching properties

Leaching tests are designed to simulate the behaviour of a material when it comes into contact with groundwater or landfill leachate. Leaching at a range of liquid to solid ratios can be combined with information on groundwater flows to predict the leaching over time of chlorides, heavy metals and other species from the residues when deposited in the ground. Tests undertaken at different pHs allow the modelling of leachate production under worst-case and normal scenarios. These results can then be used to determine the environmental impact of different residue management options.

As stated above, leaching tests are also used as part of the Waste Acceptance Criteria to determine whether a particular waste can be landfilled.

Lewin et al. (1996) used two leaching tests to assess the bottom ash samples (van der Sloot et al., 1994).

■ the maximum availability test (MALT) – a two-stage test leaching a ground sample for three hours in deionised water with the addition of nitric acid to maintain a pH of 7, followed by three hours leaching in nitric acid at a constant pH of 4. This test is designed to represent a worst-case situation.

■ the granular leach test (GLT) – a two stage test using water for 8 hours and 16 hours respectively with no pH control and using an unground sample.

The results of the MALT test showed that, under worst-case conditions, 100% of the chloride, sulphate and cadmium would be released and 6% of the lead and 1% of the chromium. The results of the GLT test (which is expected to perform in a similar way to the test required under the forthcoming WAC) suggested that the material had sufficiently low leachability in terms of chloride, sulphate, lead, mercury and cadmium to be landfilled.

Classification

In the European Waste Catalogue, incinerator bottom ash is classed as a non-hazardous waste and, as stated above, appears to meet the Waste Acceptance Criteria.

Uses

According to an Environment Agency survey carried out in 2002, about 60% of incinerator bottom ash (IBA) is sent directly to landfill and 40% to ash processing companies. The ash sent for processing is generally aged in stockpiles to allow the alkalinity to be reduced, metal oxidation to complete and the material to become more physically and chemically stable. It is then screened to reduce oversized material and the ferrous and non-ferrous metals are removed. Finally, the ash is screened into different product grades.

The final destinations of the processed bottom ash are shown in Table 18.

Table 18   Incinerator bottom ash management in the UK

Use	Percentage
Landfill disposal	38
Construction (road sub-bases, foundations and landfill engineering)	36
Stockpiled	15
Constructional block manufacture	6
Asphalt component	1

Source: Data from Environment Agency (2002).

Recent projects that have used bottom ash include (Ballast Phoenix, 2004):

- the Channel Tunnel rail link
- the Birmingham relief road (M6)
- Stansted Airport
- several roads (in cement and bitumen-bound uses).

Some other European countries have a much more advanced IBA recycling infrastructure with the Netherlands and Denmark recycling 100% and 75% of their IBA respectively. France and Germany both recycle about 50% of their IBA.

5.1.5   Air pollution control residue management

Physical and chemical characteristics

Air pollution control (APC) residues consist of a fine dusty material mainly in the size range 5–75 μm. The elemental composition is shown in Table 19.

Table 19   Air pollution control residue composition

	Units	Value
pH		12.1–12.3
Moisture	%	2.2–2.6
Loss on ignition	% (dry)	4.1–4.3
Carbon	% (dry)	2.9
Si	% (dry)	3.29
Al	% (dry)	0.6–0.9
Mg	% (dry)	0.36–0.48
Fe	% (dry)	0.17
Ca	% (dry)	34.9–35.5
Cl	% (dry)	14.7
SO_4	% (dry)	8.3
F	% (dry)	0.13
Zn	ppm (dry)	7070–10 362
Pb	ppm (dry)	2900–6548
Cu	ppm (dry)	301–409
Mn	ppm (dry)	514
Cr	ppm (dry)	143–148
Cd	ppm (dry)	82–127
As	ppm (dry)	32
Mo	ppm (dry)	<10
Hg	ppm (dry)	26–78

Source: Data from Lewin et al. (1996).

SAQ 17

Consider Tables 2, 17 and 19 and list the elements that are concentrated in bottom ash and APC residue in comparison with the raw refuse. Why is this so?

Leaching properties

APC residue is much more soluble than the bottom ash and Lewin et al.'s results suggest that it would fail to meet the Waste Acceptance Criteria leaching limits for chloride, lead and mercury.

Classification

Under the European Waste Catalogue, APC residues are classed as 'Hazardous Wastes' – mainly due to the irritant nature of the unreacted lime present. In addition, as shown above, the material will not be permitted to landfill once the WAC are implemented.

Management

At the time of writing (2006) most APC residues are landfilled at hazardous sites either in dry form or after mixing with water to form a solid mass. About 12% of the total is sent to mixed waste treatment plants to be used in neutralising acidic wastes. APC residues are also being deposited in a worked out salt mine in Cheshire along with other dry hazardous wastes. A short video clip of a typical APC landfill site is included on the Resources DVD.

However, under the WAC regulations discussed above, the landfilling of untreated residue is unlikely to continue. There are a number of processes (Lewin et al., 1995, Chartered Institution of Wastes Management, 2003) undergoing development that are designed to result in a more stable product that would meet the WAC.

Washing

One of the principle problems with APC residues is the high content of soluble chlorides, so washing is a common first stage in many treatment processes. The process equipment is relatively straightforward and consists of a washing reactor followed by a filter system. This process generates a secondary pollutant stream, namely the liquid effluent which has a high pH and chloride content and contains the soluble heavy metals. This liquid needs to be treated – usually by pH adjustment, to convert the metal salts to insoluble hydroxides, followed by filtration to remove the insoluble precipitate. The final saline effluent then needs to be either discharged to sewer (if permitted) or transported by tanker to a sewage treatment works that discharges into an estuary.

SAQ 18

An APC washing process results in the production of concentrated brine. What problems might be encountered in finding beneficial markets for this material?

Solidification

Solidification consists of encapsulating the insoluble residue from an APC washing process with cement to form a monolithic material that reduces the porosity and hydraulic conductivity of the material and hence its leachability. The process also chemically combines the potential pollutants within the crystal structure in the form of metal hydroxides or carbonates. As stated above, chlorides are not entrapped in these processes, so a preliminary washing stage may be required.

This process would normally be carried out at the disposal site rather than at the individual incinerators. Typically, this requires between one third and half a tonne of cement per tonne of APC residue.

This process is used in Germany and Switzerland. In Germany the solidified material is used as backfill material in worked-out salt mines. In Switzerland, the washed and solidified blocks are usually deposited in surface level landfills. Note that for strategic reasons, Switzerland does not permit subsurface landfill of any wastes.

Stabilisation

Stabilisation is a related process and consists of adding reagents to the residue that react with the soluble hazardous components, making them less soluble.

Thermal treatment

APC residues can be treated in high temperature processes, resulting in the formation of an inert glassy material. In vitrification processes, the residue is mixed with glass-forming compounds and heated to 1300–1500 °C. The final product is a single phase amorphous material. In the related melting process, glass-forming compounds are not added and the product is a multi-phase material. Both processes have the advantage that the volume of residue is reduced by 30–50% and the dioxins are destroyed in the high temperature process.

These processes do suffer from a number of drawbacks:

- chlorides are still a problem and pre-washing may be required
- volatile heavy metals such as mercury, lead and zinc are released on heating so abatement equipment is necessary which, in turn, generates its own residues
- the process is expensive and energy intensive.

This technology is used in Japan where around thirty such plants are operating. One installation has operated since 1977 in Europe at the CENON plant in Bordeaux, which burns MSW and sewage sludge. The vitrified residue is used in block making.

5.2   Health impacts of incineration

5.2.1   Public health impacts

DEFRA's (2004b) review of the environmental and health effects of waste management referred to in Section 4.2.2 above also considered these impacts in relation to MSW incineration.

DEFRA identified 23 epidemiological studies on the health effects of incineration that were considered to be both scientifically rigorous and relevant. However, it should be noted that many of these studies were undertaken using data relating to the operation of pre-1990 incinerators which were not fitted with any acid gas scrubbing equipment and particulate capture was limited to electrostatic precipitators. Therefore some of the reviewed research relates to much higher emissions than from present-day incinerators.

DEFRA also identified a number of other problems with these past studies:

- socio-economic background of the people studies were a major confounding factor
- the impacts of other local sources of pollutants were often neglected
- several of the studies that considered distance of residents from the source neglected the effects of prevailing winds
- the causes of cancer-related deaths were often attributed to secondary tumours rather than the primary tumour which ultimately caused the death.

In spite of these problems, DEFRA reviewed data relating to 14 million people. In terms of cancer they concluded that:

> [T]here is no consistent evidence of a link between exposure to emissions from incinerators and an increased rate of cancer.

They went on to cite the Committee on the Carcinogenicity of Chemicals in Food, Consumer Products and the Environment's conclusion that:

> [A]ny potential risk of cancer due to residency (for periods in excess of ten years) near to municipal solid waste incinerators was exceedingly low and probably not measurable by the most modern techniques.

On the question of respiratory problems, the report concluded:

> [T]here is little evidence that emissions from incinerators make respiratory problems worse. In most cases the incinerator contributes only a small proportion to the local level of pollutants.

The effect of incineration (particularly in relation to dioxins and related compounds) on the reproductive system has received much publicity in recent years. DEFRA reviewed research on the incidence of twin births, effects on sex ratios, kidney function, sexual development and congenital malformations. This review found that the studies either:

- reported no adverse effects
- identified adverse effects but were unable to establish a causal link (i.e. an association was found, but could not be explained), or
- produced results that should be negated due to changes in medical practices over the study period.

These largely negative findings are not surprising. In Section 4.1.1, I showed that the emissions from a modern incinerator only account for a small percentage of the background levels. Furthermore incineration accounts for a relatively small proportion of the UK's atmospheric pollution inventory, as shown in Table 20.

Table 20   Contribution of incineration to UK pollutant emissions

Pollutant	Contribution of MSW incineration to UK total (%)
Carbon dioxide	1.6
Methane	0.001
Particulate matter	0.05
NO_x	0.26
SO_2	0.009
HCl	0.17
Dioxins and furans	0.2
Cadmium	0.2
Mercury	1.4

Source: Adapted from DEFRA (2004b).

MSW incineration is often perceived as a major (if not the principal) source of dioxins, but Table 20 shows that this is not the case. Table 21 lists the UK's main sources of dioxins.

Table 21   UK dioxin sources

Source	Emission (g y^{-1} TEQ)
Domestic fires	65
Iron and steel industry	61
Fireworks	50
Non-ferrous metal industry	22
Power generation	14
Road transport	12
Cement manufacture	2.3
Landfill	1.9
Chemical waste incineration	1.3
MSW incineration	0.97
Clinical waste incineration	0.4
Others (other industry, accidental fires, small-scale waste burning – building sites and bonfires, etc.)	130

Source: DEFRA (2004b).

TEQ, toxic equivalent.

However, there is no room for complacency. Under the integrated pollution prevention and control (IPPC) regime (discussed in the Legislation Supplement), operators must use the best available techniques to eliminate or minimise pollution. As pollution technology improves, there should be a corresponding reduction in emissions from all processes governed by IPPC. For incineration this means not only the atmospheric emissions, but also the impact of the incineration residues and local impacts such as noise, odours, fugitive emissions and visual impacts.

5.2.2   Health impacts on site operators

The impact on the health of incinerator operators also needs to be considered. Gustavsson (1989) studied an incinerator in Sweden that closed in 1986. Dust levels in the incinerator were extremely high: 82 mg m^{-3} compared with the current workplace exposure limit of 10 mg m^{-3}. Not surprisingly, Gustavsson found that deaths due to lung cancer were twice the local rate.

In a similar study from a more modern installation, Rapiti et al. (1997) concluded that the mortality rate was lower than expected and, taken as a whole, the incidence of cancer was comparable to those in the local community. In terms of types of cancer, incinerator operators had lower rates of lung cancer, but higher incidence of gastric cancer.

6   LANDFILL

6.1   Introduction

> **READ**
>
> At this point, you should read the material on the Landfill Directive given in the Legislation Supplement and in Sections 2.2.3, 8.3 and 15.1 of T210 Block 4: *Wastes management*.

To comply with the requirements of the Landfill Directive and national recycling/recovery targets, the UK will be required to reduce the amount of municipal solid waste (MSW) disposed of to landfill. However, there will still be a need to landfill MSW that cannot be recovered without excessive environmental and/or financial costs. Also, once landfilled, MSW remains active for many decades so today's landfill sites will require to be monitored and the emissions controlled until the end of the century and possibly beyond.

Several methods are being used to reduce the amount of biodegradable municipal solid waste (BMSW) disposed of to landfill. Incineration, composting and recycling are the most obvious, but novel techniques such as mechanical biological pre-treatment (MBP) are under active consideration. All these processes generate residues that are eventually landfilled. These residues all have the potential to cause environmental pollution in landfills, but these potential pollutants will differ from process to process and will certainly cause different problems compared with untreated MSW when landfilled.

In this section we will consider the processes taking place in landfill sites (which should be read in conjunction with Section 8.1 of T210 Block 4: *Wastes management*). I will then consider ways of modelling the formation of landfill gas and leachate and assessing their environmental impacts including groundwater protection risk assessments. Finally, the implications these have on the design and operation of landfill sites will be considered.

6.2   Landfill regulation

The design, construction, operation and post-closure aftercare of landfilling are regulated under the pollution prevention and control (PPC) regime. The Landfill Regulations (which implement the Landfill Directive) were introduced in 2002 and supplement and amend the PPC regulations in relation to landfill sites. Collectively, these regulations are concerned with protecting the environment from the impacts of landfill and are summarised below.

6.2.1   Landfill permits

All landfill sites require a permit from the national regulator. These permits govern the operation of the site, the types of waste accepted, monitoring and reporting, and the measures taken to prevent harm to the environment and human health. These permits replace the waste management licences issued under the Environmental Protection Act 1990. All new sites require a permit before they can begin operating and existing sites have also had to go through the process of obtaining permits.

In addition, all landfill sites must have planning permission from the appropriate local authority. As well as meeting normal planning requirements, permission will only be granted if the requirements relating to site location shown below are met.

6.2.2   Site location

The following factors should be taken into consideration when siting landfills:

- the distance from the site boundary to residential and recreational areas, waterways, water bodies and other agricultural or urban sites
- groundwater, coastal waters or nature protection zones in the area
- local geological and hydrogeological conditions
- the risk of flooding, subsidence, landslides or avalanches
- the protection of the natural or cultural heritage of the area.

6.2.3   Site classification

All landfills will be classified in their permits as accepting either hazardous, non-hazardous or inert wastes.

6.2.4   Prohibitions of certain wastes

No landfills will be permitted to accept:

- liquid wastes
- explosive, corrosive, oxidising, flammable or highly flammable materials
- tyres (other than bicycle tyres and those with a diameter above 1400 mm).

In general, all wastes must be pretreated before landfilling unless:

- the waste is inert and treatment is not technically feasible, or
- treatment would not reduce the quantity for landfill or the hazards it poses to human health and the environment.

6.2.5   Groundwater protection

The Environment Agency's policy is to assess potential site locations with respect to groundwater protection on a site-by-site basis. However it will always object to applications for sites based in a groundwater 'Source Protection Zone I'. In addition, if the risk assessment indicates that active long-term site management is necessary, the Agency will oppose applications for sites in the following areas:

- above a 'major aquifer'
- within Source Protection Zones II and III
- below the water table in any strata where the groundwater provides an important contribution to river flow or other sensitive surface waters.

Groundwater Source Protection Zones are areas identified by the Environment Agency as requiring special protection from groundwater pollution due to their proximity to important sources of water. Zone I is where the travel time of water to the abstraction point is less than 50 days, Zone II is where the travel time is less than 400 days and Zone III covers the remainder of the catchment zone.

6.2.6   Water control and leachate management

The directive specifies that appropriate measures should be taken to:

- control rainwater and prevent surface and groundwater from entering the landfilled waste
- collect leachate and contaminated water and treat to the appropriate discharge standard.

The Regulations specify the structure and properties of the liner and capping layers for each type of landfill site, as summarised in Tables 22–4.

Table 22   Landfill base and side requirements

Site category	Mineral layer thickness (m)	Permeability (m s^{-1})
Hazardous waste	$\geqslant 5$	$\leqslant 10^{-9}$
Non-hazardous waste	$\geqslant 1$	$\leqslant 10^{-9}$
Inert waste	$\geqslant 1$	$\leqslant 10^{-7}$

Note that alternative liners, having equivalent or better performance to a liner with the above specifications, are acceptable.

Table 23   Landfill leachate collection and sealing system

Parameter	Hazardous waste sites	Non-hazardous waste sites
Artificial sealing layer	required	required
Leachate drainage layer $\geqslant 0.5$ m	required	required

Note that the artificial sealing layer is in addition to the mineral layer specified above.

Table 24   Landfill capping requirements

Parameter	Hazardous waste sites	Non-hazardous waste sites
Gas drainage layer	required	not required
Artificial sealing layer	not required	required
Impermeable mineral layer	required	required
Drainage layer >0.5 m	required	required
Top soil cover >1 m	required	required

6.2.7   Landfill gas

All sites that accept biodegradable wastes must be fitted with systems to collect the landfill gas formed. Where possible the gas should be used to produce energy and where this is not possible the gas must be flared. In either event, this must be done in a way that minimises damage to the environment and risks to human health.

6.3   Landfill processes

The chemical, physical and biological processes taking place in landfills are summarised in Section 1.2.3 above and discussed in detail in Section 8.1 of T210 Block 4: *Wastes management.* At this point, you should make sure that you are familiar with these processes and their timescales.

6.3.1   Landfill gas formation and composition

Although some gaseous products are formed during the early stages of the decomposition process, the bulk of the landfill gas is formed during the methanogenic phase. This usually begins some 6–12 months after the sealing of the cell and can continue for many years.

The main components are methane and carbon dioxide formed by the breakdown of fatty acids (principally acetic acid)

$$2CH_3COOH \longrightarrow 2CH_4 + 2CO_2$$

Methane is also formed by the microbial reduction of carbon dioxide

$$4H_2 + CO_2 \longrightarrow CH_4 + 2H_2O$$

The concentration of methane in the landfill gas generally has a maximum value of around 65%. When the methane has reached its maximum concentration, the evolution of the methane/carbon dioxide mixture will continue for many years at a steady rate, before it begins to decline. It can be expected that in large, deep landfill sites, methane will be produced at maximum concentrations for periods in excess of ten years. Figure 6 shows the changes of gas composition with time. The total length of time during which methane is produced will vary from site to site, but this stage can last between 20 and 50 years.

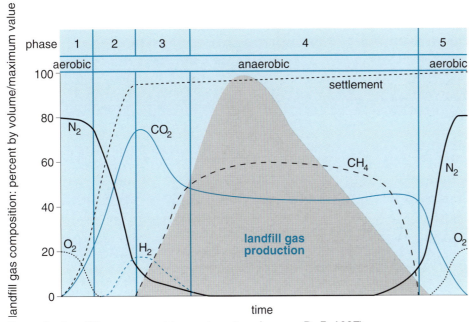

Figure 6   Landfill gas composition against time (source: DoE, 1997)

The change in composition of domestic waste over the years has had a considerable effect on landfill gas production. Until the early 1960s, domestic waste had a much lower biodegradable content than today's waste. For example, 50% was coal ash from domestic fires. This has all but disappeared, and has been replaced by unburned materials (paper, kitchen waste, plastics and so on). There has thus been a change from a large proportion of relatively inactive material to much more organic and biodegradable material, which has resulted in the production of more gas.

The operation of sites has also changed. Current techniques reduce air ingress by such means as:

- capping sites on completion
- using impermeable intermediate cover
- employing techniques which produce higher compaction of the waste.

All this helps to promote the anaerobic conditions necessary for the production of methane. The generation of landfill gas is also affected by:

- *water content of the waste*. Increased moisture content enhances decomposition and hence gas production. The optimum conditions for anaerobic digestion normally occur in unsaturated wastes with a moisture content of more than 40%

■ *density.* Lower densities allow easier water penetration and this stimulates a more rapid reaction

■ *pH.* Methane-forming bacteria can survive only within a narrow pH range, the optimum for methane production being between 6.4 and 7.4

■ *temperature.* The optimum for anaerobic digestion in landfills occurs between 29 °C and 37 °C.

Although landfill gas consists principally of methane and carbon dioxide, there are also small amounts of other gases, including hydrogen, hydrogen sulphide, and a wide range of alkenes, alcohols, esters, halocarbons, organosulphur compounds, aromatic compounds, terpenes and other organic materials.

Hydrogen sulphide concentration is usually low (<10 ppm) but it *can* occur in significantly higher concentrations, for example if large quantities of materials containing a high proportion of sulphate, such as gypsum waste, are mixed with domestic and commercial waste. The characteristic odour of landfill gas is due to the presence of trace constituents such as mercaptans. Table 25 gives a typical analysis of the gas produced in the steady state.

Table 25   Landfill gas composition

Component	Units	Range of values
Methane	%	44–59
CO_2	%	33–52
Oxygen	%	<0.01–3
Hydrogen	%	<0.01–0.08
Moisture	%	1–4
Total S	mg Nm^{-3}	31–430
Total F	mg Nm^{-3}	<0.6–20
Total Cl	mg Nm^{-3}	<0.6–78
H_2S	mg Nm^{-3}	21–400
CO	mg Nm^{-3}	22–146
Non-methane volatile organic compounds	mg Nm^{-3}	<120–1440

Source: Data from Gregory et al. (2003).

6.3.2   Landfill gas production and release modelling

As part of its work on greenhouse gas emissions, the Intergovernmental Committee on Climate Change (ICCC) is involved in producing estimates of the amount of methane released from landfill sites. It developed a methodology for modelling landfill gas emissions which has been refined by AEA Technology (Brown et al., 1999). This model assumes that the formation of landfill gas can be simulated by means of a 'first order exponential decay' model. Such models are based on the assumption that the rate of a reaction is proportional to the amount of reactant present. The general form of the model is

$$C = C_0 e^{-kt}$$

where

C is the amount of reactant remaining after time t

C_0 is the amount of reactant present at the start ($t = 0$)

k is the decay constant for the reaction (dimensions $time^{-1}$).

In the case of landfill gas, the equation can be modified to give the amount of methane released in a given year after landfilling using the following equation

$$Q_T = L_T(e^{-kT}(1 - e^{-k}))$$ (1)

where

Q_T is the amount of methane released in year T (tonnes)

k is the rate constant (y^{-1})

and

$$L_T = DOC \times z \times r_{mc} \times \frac{16}{12}$$ (2)

where

DOC is the amount of degradable organic carbon in the original waste (tonnes)

z is the fraction of the DOC that degrades in the landfill

r_{mc} is the proportion of the degrading carbon that forms methane rather than carbon dioxide

the fraction $\frac{16}{12}$ allows for the mass conversion of carbon C to methane (CH_4).

AEA Technology assumed that degradable waste can be classified into three classes each with its own decay constant and value for r_{mc}:

■ readily degradable (RDO) – consisting typically of kitchen and non-woody garden wastes

■ moderately degradable (MDO) – consisting typically of paper

■ slowly degradable (SDO) – consisting typically of plastics.

Equation 1 can be used separately for each fraction.

COMPUTER ACTIVITY 1

Use Equation 1 and the following data to produce a spreadsheet and graph showing the amount of methane formed (from each class of waste and the total) during the first 30 years after the landfilling of 100 000 tonnes of waste.

Category	Quantity (t)	Rate constant k (y^{-1})	r_{mc}	z
RDO	20 000	0.185	0.5	0.75
MDO	35 000	0.1	0.5	0.5
SDO	8 000	0.05	0.5	0.3

SAQ 19

What do the model and graph produced for Computer Activity 1 tell us about the amounts of methane produced, the timescale and the importance of each fraction of material?

The model shown in Computer Activity 2 develops the use of this equation further to predict the methane generated over a 30-year period by a landfill that takes waste over a 10-year period. The model also calculates the amount of methane released to the environment after taking account of the proportion captured by the gas control system and the amount oxidised by microbial action in the landfill capping material.

COMPUTER ACTIVITY 2

Use the model given in this activity on the Resources DVD to investigate the effectiveness of the following landfill gas control measures on the amount of methane released and the timescales for these releases:

■ improving methane capture efficiency

■ reducing the input of readily degradable waste with a corresponding increase in slowly degradable waste

■ reducing the input of readily degradable waste with no increase in the other types of waste.

6.3.3   Landfill gas control and energy recovery

Section 6.3.1 discussed the formation and composition of landfill gas. In this section its control and utilisation are considered.

Landfill gas, which can be produced for up to fifty years after waste is deposited and can migrate long distances from the site, is potentially hazardous in a number of ways. Appropriate action must therefore be taken to minimise the risks, both in the design stage and in the operation of landfill sites. Wastes Management Paper 26 (DoE, 1997) puts the problems into the following categories:

(a) explosions or fires due to gas collecting in confined spaces, such as buildings, culverts, manholes or ducts on or near landfill sites;

(b) asphyxiation of people entering culverts, trenches or manholes on landfill sites;

(c) the risk of waste being set on fire, following ignition of landfill gas when it is released through fissures in the surface;

(d) detrimental effects on crops or vegetation on or adjacent to landfill sites;

(e) risks to human health from gas emissions;

(f) nuisance problems, especially odour.

The Landfill Directive (Council of the European Union, 1999) recognises the environmental and other potential risks arising from landfill gas and states that:

> Appropriate measures shall be taken in order to control the accumulation and migration of landfill gas.

> Landfill gas shall be collected from all landfills receiving biodegradable waste and the landfill gas must be treated and used. If the gas collected cannot be used to produce energy, it must be flared.

> The collection, treatment and use of landfill gas shall be carried on in a manner which minimises damage to or deterioration of the environment and risk to human health.

In addition to the risks of explosion described above, damage to vegetation can arise from landfill gas because the methane in the gas displaces the normal soil atmosphere. This prevents the diffusion of oxygen from the air into the soil, which then becomes anoxic. Phytotoxic compounds are also usually present in the gas and stunted growth results, particularly if roots penetrate the landfill site cap.

Where sites are surrounded by medium-to-high-permeability strata such as sands, gravels, chalk or fissured rock, the gas can move laterally for long distances beyond the site boundary. Generally it will vent to atmosphere, but, as seen above, it can collect in buildings and so create potential fire and explosion hazards as well as a nuisance from smell. Figure 7 shows possible gas migration paths.

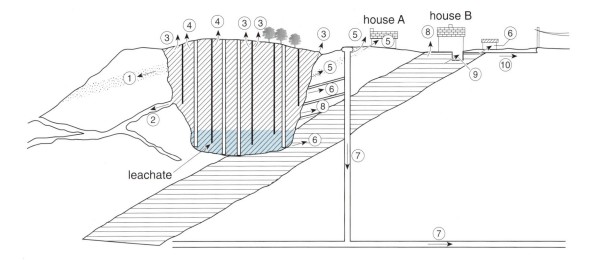

Gas pathways to atmosphere

① through high permeability strata down the bedding plane

② through caves/cavities

③ through desiccation cracks of the capping at the site perimeter, around tree roots, etc.

④ around site features which provide vertical pathways, gas wells or leachate wells

⑤ through high permeability strata up the bedding plane, to atmosphere or house A

⑥ through fissures caused by explosives etc.

⑦ along man-made shafts etc.

⑧ through highly fissured strata into the atmosphere or buildings such as house B or shed etc.

⑨ into underground rooms

⑩ along underground services

Notes

(a) Gas may vary depending on its source from within the landfill and the migration route, e.g. route 5 gas compared with route 3 gas

(b) Leachate may degrade to give rise to gas generation at some distance from the site

Figure 7   Possible landfill gas migration paths from a completed/restored site (source: DoE, 1989)

Lateral migration of the gas will be prevented by the installation of the impermeable barriers required to prevent leachate migration. However, the combination of these barriers and the impermeable cap mean that a gas extraction system is essential if biodegradable waste is present in the landfill.

Gas extraction systems

Preliminary assessment of the quality and quantity of gas is an essential first step in any recovery scheme. Test wells will need to be drilled and pumping tests done. These should assess not only the quantities of gas likely to be produced but also the sphere of influence of the well, which may be as much as 30–50 m.

Theoretically, some 400 m^3 of gas can be produced from each tonne of waste. In practice, actual amounts produced are considerably less than the theoretical value. In terms of rate of production, the Environment Agency/SEPA (2002)

suggest that small sites containing around 100 000 m³ of MSW will produce gas at the rate of 25–100 m³ h⁻¹. For sites containing 1–10 million m³ the gas production rates will range from 250 to 10 000 m³ h⁻¹.

Extraction wells usually have a diameter of 0.3–0.6 m. Wells installed as the site is filled generally extend to the base of the site, whereas wells that are drilled after filling extend to 70% of the depth of the site (to avoid the risk of damage to the liner). The well casing is approximately 0.1 m in diameter and is perforated for the bottom half, the annulus around the lower part being backfilled with permeable material such as gravel, rock or broken bricks. This is then capped with clay or other sealing material. The head of each well may be fitted with a flow/pressure control valve and suitable monitoring points for flow rates, and measurements for temperature and pressure.

Individual wells are joined together with plastic pipe of at least 100 mm diameter. Suitable materials include uPVC, HDPE or polypropylene. The pipes can either be laid on the surface of the site or buried, and must be sloped to facilitate collection and removal of condensate. The pipe system is connected to a suitable gas blower or compressor. Care must be taken in choosing the type of gas mover as some types are not suitable for use with landfill gas because of their sealing arrangements.

Any scheme to recover energy must be operated so that the gas is extracted at a sufficient rate to eliminate potential hazards and nuisance. Ideally, therefore, the energy load pattern should match the extraction rate. However, in practice, schemes need either a means of flaring excess gas or storage capacity to buffer the changing load pattern.

If the gas is to be flared, it is essential that a flame trap or water seal trap is installed between the pump and the flare stack. Flares should be installed at a high level to limit the nuisance from emissions. A gas stream monitor or flame temperature sensor should also be installed and connected to an alarm-and-call-out system, to ensure rapid response to failure. The flare may require a continuous ignition source, fuelled, for example, by liquefied petroleum gas (LPG).

Condensate knockout is a necessary part of a gas recovery system. This is because the gas leaving the wellhead is saturated and at a temperature of approximately 35 °C, and subsequent cooling to ambient temperatures produces condensation. In addition, foam can be entrained in the gas stream because of the low pressure and relatively high velocity in the gas well. Foam is eliminated using coalescing meshes, and both the foam and the condensate can be collected in an expansion vessel. Plate-type towers are an alternative means of removal.

Gas quality should be monitored continuously. Oxygen levels can be monitored using a paramagnetic measuring cell. Limits should be set at less than 5% oxygen by volume and these should be linked to the compressor motor, to shut down the plant in the event of higher oxygen levels being detected.

Because of the dangerous nature of landfill gas, it is essential that equipment used is designed and built to the highest standards, and the use of makeshift equipment is unacceptable. Gas leakage or excessive air ingress into the plant or landfill site must be avoided, and necessary safety precautions, such as pressure testing of the pipework and installation of pressure relief devices, must be taken.

In a recovery plant there may also be additional oxygen and nitrogen as a result of air being drawn into the system. This can occur either in the site itself or from leaks or fractures in the pipeline or elsewhere in the plant.

When air is mixed with landfill gas, the mixture can become explosive. The flammable range (also called the explosive limits) for pure methane is between

5% and 15% by volume, but these limits are affected by the presence of carbon dioxide. Carbon dioxide has a specific gravity of 1.5, compared with 0.5 for methane. As a result of this, landfill gas may be more dense or less dense than air, depending on the composition.

6.3.4   Landfill gas combustion

Ideally, landfill gas should be burned in systems which incorporate energy recovery. The remote nature of most landfills means that local markets for heat or steam are limited, so the gas is generally used in engines to generate power.

The ranges of emissions from nine landfill gas-fired engines are shown in Table 26.

Table 26   Landfill gas engine emissions

Emission	Units	Range of values
CO_2	%	13.1–17.6
CO	mg m^{-3}	508–2600
SO_2	mg m^{-3}	18–540
NO_x	mg m^{-3}	360–1500
HCl	mg m^{-3}	0.2–584
Particulate material	mg m^{-3}	1.7–51
Total hydrocarbons	mg m^{-3}	530–5260
Dioxins	ng m^{-3} (TCDD equivalent)	0.0009–0.61

Source: Data from Gregory et al. (2003).

However, not all sites produce sufficient gas to justify the cost of gas engines and grid connections. In these cases flares are used. Flares are also used on sites with gas engines as stand-by equipment or to cope with periods of excessive gas release.

The Environment Agency/SEPA (2002) have suggested emission limits for gas flares, presented in Table 27.

Table 27   Landfill gas flare emission limits

	Emission limit mg Nm^{-3} (corrected to 3% oxygen)
CO	50
NO_x	150
Unburned hydrocarbons	10

SAQ 20

A landfill gas engine discharges 30 Nm^3 s^{-1} of combustion gas with an effective discharge height of 40 m. The engine is located 400 m from the site boundary. Using the point source dispersion model in the Air section of the Resources DVD computer activities, estimate the ground level concentration of NO_x at the boundary due to the operation of the engine for the emission values reported in Table 26 and for the Environment Agency's proposed limit of 250 mg Nm^{-3}.

6.3.5   Leachate formation and composition

Leachate is formed by rain and surface water percolating through the landfilled waste. As it passes through the waste, soluble solid compounds in the waste dissolve in the water, and the liquids formed during the decomposition of the waste or present in the original waste also accumulate in the leachate.

Consequently, the composition of leachate depends on the nature of the landfilled waste and on the stage in the decomposition processes. The characteristics of MSW landfill leachate formed in the acetogenic and methanogenic phases are summarised in Table 28 and typical values are given in Table 29. Leachates produced in the methane phase are often referred to as being stabilised. Less than 20% of the organic carbon in stabilised leachates is present as fatty acids because these readily soluble and degradable components have been lost over time. This is indicated by the low BOD:COD ratio.

Since landfills operate over many years and contain wastes of different ages, in practice the characteristics of leachates fall between the two extremes shown in Table 28. It is also worth noting that, while dilution by rainwater would reduce the concentrations of all the compounds present in the leachate, the ratios of these concentrations would remain the same. Hence, such ratios are useful indicators of leachate characteristics.

Table 28   Characteristics of MSW landfill leachate

Acetogenic phase	Methanogenic phase
High fatty acid concentration	Very low fatty acid concentration
Acidic pH	Neutral to alkaline pH
High BOD (often >20 000 mg l^{-1})	Low BOD (>200 mg l^{-1})
COD of several thousand mg l^{-1}	COD of several hundred mg l^{-1}
High BOD:COD ratio	Low BOD:COD ratio
Several hundred mg l^{-1} of ammonia	Several hundred mg l^{-1} of ammonia
Several hundred mg l^{-1} of organic nitrogen	
Strong unpleasant smell	
Possible high concentrations of iron, manganese, calcium, magnesium and moderately high concentrations of some heavy metals, e.g. zinc	Low concentrations of metals
High concentrations of other inorganics, e.g. Na, K, Cl	High concentrations of Na and Cl, moderately high concentrations of other soluble inorganics

Source: Adapted from Knox (1985).

Table 29   Comparison of leachates from landfills in the acid phase and methanogenic phase of stabilisation

	Acetogenic phase		Methanogenic phase	
	UK	Europe range	UK	Europe range
pH	6.35	4.5–7.8	7.90	6.8–9
total organic carbon (TOC)	18 600	1010–29 000	260	184–2270
chemical oxygen demand (COD)	59 200	6000–15 2000	804	500–8000
biochemical oxygen demand (BOD)	35 000	4000–68 000	18	20–1770
Ammonia nitrogen (N as NH_3)	880		440	
organic nitrogen (as N)	600		8.3	
oxidised nitrogen (as N)	2.0		21	
orthophosphate (as P)	1.4		0.4	
total suspended solids (105 °C)	221		61	
Conductivity ($\mu S\ cm^{-1}$)	–		11 600	
Sodium	1800		2220	
Potassium	1650		515	
Calcium	5050	10–6240	232	20–600
Magnesium	400	25–1150	224	40–478
Chloride	2645		3000	
Sulphate (as SO_4)	–	<5–1750	460	<5–420
Zinc	200	0.1–140	0.18	0.03–6.7
Copper	0.04		0.09	
Nickel	1.8		0.12	
Cadmium	0.012		<0.02	
Lead	0.44		<0.1	
Chromium	1.5		0.08	
Iron	1470	20–2300	3.8	1.6–280
Manganese	230	0.3–164	0.41	0.003–45

Sources: Data from (Europe) Vasel et al. (2003); (UK) Robinson (1995).

All concentrations in mg l^{-1} (except pH and conductivity).

6.3.6   Leachate from landfilled incineration residues

The leachates discussed above are derived from landfilled raw waste. However, as the requirements of the Landfill Directive begin to be implemented, the amount of raw waste landfilled will be reduced and there will be an increase in the landfill of incineration residues. The leachate from incinerator residues has lower BOD and COD values but much higher levels of salts. Table 30 shows the composition of leachate from an ash landfill site in Japan.

Table 30   Ash landfill leachate

Parameter	Units	Concentration
pH		7.3
Conductivity	mS m^{-1}	4540
Total hardness	mg l^{-1} as $CaCO_3$	17 900
Cl	mg l^{-1}	26 300
Ca	mg l^{-1}	7120
Na	mg l^{-1}	5590
K	mg l^{-1}	5280
BOD	mg l^{-1}	176
COD	mg l^{-1}	226

Source: Data from Ushikoshi et al. (2001).

Such a leachate would not be amenable to treatment at a conventional sewage treatment works, and in this case it is treated in a reverse osmosis system before discharge.

6.3.7   Water balance calculations

One particularly important consideration in landfill site design and operation is the water balance. Water balance calculations are used both in assessing the degree of pollution potential and in planning operational aspects, in order to allow the leachate to be dealt with satisfactorily. The two basic questions to be answered are:

■   How long will it take before a landfill site begins to generate leachate?

■   How much leachate will be generated per year once leachate generation has begun?

Many different equations for performing water balance calculations appear in the literature but they all follow the same basic principle, that of *mass balance*. The various possible liquid inputs and outputs to a site are given in Table 31.

Table 31   Possibilities for liquid inputs and outputs

Inputs		Outputs	
Precipitation	P	Evaporation	E
Surface water run-on	R_{on}	Transpiration	T
Groundwater infiltration	I	Surface water run-off	R_{off}
Leachate added	L_{in}	Leachate removed	L_{off}
		Surface seepage	L_{sep}
		Groundwater recharge	L_{rec}

SAQ 21

Which of the inputs and outputs given in Table 31 are potential pollutants?

The rate of inflow minus the rate of outflow is the rate of change in the quantity of liquid, S, stored within the site, and is given by

$$S = (P + R_{on} + I + L_{in}) - (E + T + R_{off} + L_{off} + L_{sep} + L_{rec})$$

If the total inputs exceed the total outputs, the quantity of liquid in the site will increase until it reaches the field capacity (see below). Note that all the above quantities with the symbol L represent leachate. In some cases the surface run-off R_{off}, which is normally clean water, can become contaminated and will need to be collected and treated.

Although the principle is straightforward, the application of water balance calculations is often difficult in practice because of uncertainties in assigning values to the variables. It may be difficult, for example:

- to assess the degree of surface water run-off
- to assess groundwater infiltration
- to determine waste densities and water retention characteristics
- to assess evapotranspiration.

There are also the problems of forecasting rainfall.

As liquid from the various inputs enters the site, it will move downwards through the waste, and some will be absorbed. As the amount of liquid absorbed increases, a point will be reached where it can no longer be held by capillary attraction within the microstructure of the waste particles, against the pull of gravity. The waste is then said to be at *field capacity*. The liquid content at field capacity is difficult to establish. The value is affected by the nature and intensity of the wetting process since rapid infiltration is more likely to find rapid drainage routes. However, figures between 30% and 40% by volume may be considered typical.

The *absorptive capacity* can be defined as the amount of liquid that can be taken up by the waste, that is, the difference between the field capacity and the initial moisture content. Liquid taken up in this way does not drain, and therefore does not appear as leachate. Typical values of absorptive capacity range from 10% to 20% by volume. However, there is considerable uncertainty in the reliability of estimates.

After the field capacity has been reached, additional liquid can still be taken up and stored as free liquid in the voids between the waste particles. This additional liquid will drain if allowed to do so, and this will result in leachate seeping from the site. Eventually, all the voids will be full and the waste is then said to be *saturated*. When saturated, waste can typically contain between 60% and 70% liquid by volume. The volume of liquid required to take the waste from its field capacity to saturation is known as the *storage coefficient* (it is also called the *porosity* or *voidage*) and is typically 20–35% by volume. Once the moisture content exceeds the storage coefficient the site will not only produce leachate but it will overflow.

The initial in-place moisture content of a waste depends on the waste type, its age and the time of year. Figures commonly assumed range from 25% to 30% of the undried waste mass (33–43% of the dried waste mass) although it could be as high as 40%.

SAQ 22

Assume that the in-place density of a given waste is 0.7 t m^{-3}. What is the initial moisture content on a volume/volume basis for a moisture content of

(a)   25%

(b)   40%

of the undried mass?

SAQ 23

Draw a diagram to show the relationship between initial moisture content, field capacity, saturation, absorptive capacity and voidage.

Add typical values of each of these to your diagram.

In theory, leachate production can be prevented, provided that the refuse filling takes place at such a rate that the liquid added to the site is insufficient to allow the absorptive capacity to be reached. In practice, however, because different wastes have different characteristics, the body of the site can be very heterogeneous.

Some impermeable items cause the effect of channelling and short circuiting, thus preventing the liquid from reaching some of the absorbent material. Hence, percolation may occur before all the waste reaches its field capacity, resulting in leachate production. Intermediate cover materials, or highly compacted wastes, can equally give rise to localised areas of saturation, known as *perched water tables*.

So the actual water-retention characteristics are dependent both on waste density and the presence or absence of voids or impenetrable barriers that can inhibit continuous downward percolation of the liquid. Both the above factors lead to difficulties in making predictions.

Although the inputs and outputs change continuously, calculations are in practice carried out over a limited number of discrete time periods. These are then assembled to cover the required overall time. Any time period can be chosen, but in general the longer the period, the less accurate the results. For example, if a time period of one year were chosen, the seasonal variation in leachate production could not be predicted.

The pattern of leachate generation can vary considerably, as shown in Figure 8. A knowledge of the variations is important in the design of treatment plant.

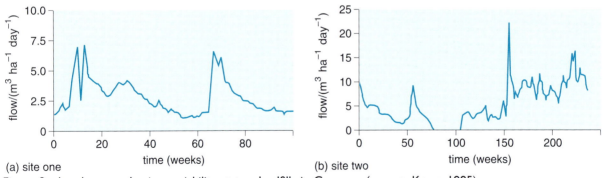

(a) site one   (b) site two

Figure 8   Leachate production variability at two landfills in Germany (source: Knox, 1985)

The terms E and T in the water balance equation are generally combined to give a term for evapotranspiration. The rate of evapotranspiration depends on:

■ intensity of the rainfall

■ nature of the cover and its initial water content

■ ambient temperature

■ soil temperature

■ humidity

■ atmospheric pressure

- wind
- slopes on the site
- sheltering conditions
- extent of vegetation.

Many of the above factors will be affected by the location of the site.

Figures for evapotranspiration of 45% for soil and over 60% for a vegetated surface have been suggested. However, conditions during summer months will often arise when evapotranspiration can reach 100%.

Another frequently used term is *effective rainfall*. This is defined as the actual rainfall minus potential evapotranspiration (PET). A figure for the effective rainfall based on measurements at selected sites can be obtained from the Meteorological Office.

The amount of liquid that actually infiltrates the site will be effective rainfall less any run-off. Experimental data suggest that infiltration rates through typical daily cover soils range from 20% in summer to 100% in winter. It appears that infiltration rate is not significantly affected by the type or quantity of intermediate cover used.

It is unlikely that run-off will be effective at an operational site. On completed sites, however, the use of contouring and drainage systems and a low-permeability cap can divert a substantial proportion of precipitation as run-off. The effectiveness of run-off is affected by the quantity, depth, compaction and surface profile of the cover material as well as by the frequency, intensity and duration of rainfall. Typically, infiltration on completed sites is 25–30% of the effective rainfall.

At many landfills, initial in-place waste densities of 0.7–0.8 t m^{-3} are achieved. At such densities, it is likely that 0.1–0.2 m^3 of added liquid per m^3 of waste as received can be absorbed before substantial leachate generation occurs. At higher compaction densities, absorptive values will fall, and at 1 t m^{-3} the absorptive capacity could be as low as 0.02–0.03 m^3 liquid per m^3 of waste as received.

Generally speaking, the higher the rate of waste input, the less likely it will be that leachate will be generated during the operational phase of the site. If the rate is so high that the absorptive capacity of the waste will not be utilised, the subsequent biodegradation and hence gas generation will be affected because there is insufficient water present.

It is good practice to operate a site to try to obtain zero free leachate. However, because of the uncertainties in parameters and variations in operating conditions and waste types involved, water balance calculations cannot be used to show conclusively that *no* leachate will occur. They can, however, be used to give an indication of the *risk* of leachate occurrence, and to help to assess the likely scale of leachate collection and treatment required.

In addition, an understanding of the principle of mass balance can help site designers and operators to reduce leachate production – for example, in the provision of surface and groundwater cut-off systems and in varying the cell size, and hence surface area, according to the waste quantities and types coming to the site.

The application of water balance calculations can be illustrated in Computer Activity 3.

COMPUTER ACTIVITY 3

The Waste Management models on the Resources DVD include one that allows you to carry out water balance calculations. Use the model to answer the following questions.

1   Consider a hypothetical landfill site cell with an area of 10 000 m² and a mean waste depth of 5 m. The initial moisture content of the waste is 20% by volume, the moisture content at field capacity is 40% by volume and the voidage is 25% by volume. The effective annual rainfall is 220 mm.

   What is the theoretical time before leachate is produced? (You should assume that leachate production begins when the absorptive capacity is reached.)

2   For the same cell, assuming that it took 5 months for the cell to be filled and that the annual rainfall is 600 mm and the cap reduces infiltration to 20% of the rainfall, calculate:

   (a) the amount of water entering the site during infilling

   (b) the total time until leachate first appears

   (c) the amount of leachate generated per year once leachate formation has started.

6.3.8   Leachate migration through site liners

In Section 8.3.1 of T210 Block 4: *Wastes management* I considered the migration of leachate through a simple mineral liner (such as compacted clay). The performance of such a liner is modelled using Darcy's law in the following equation

$$Q = k \times A \times \frac{(h+d)}{d}$$

where

   Q is the leachate flow (m³ d⁻¹)

   k is the permeability of the liner (m d⁻¹)

   A is the site or cell area (m²)

   h is the head (or depth) of leachate above the liner (m)

   d is the thickness of the liner (m).

SAQ 24

Consider a landfill site of area 2500 m² lined with 1.5 m of compacted homogeneous clay with a permeability of 10^{-9} m d⁻¹. If leachate pumping limits the head of leachate to 0.8 m, what is the annual flow of leachate from the site?

In reality, liners are often more complex. The Environment Agency (2002) has issued a guidance note that interprets the requirements of the engineering aspects of the Landfill Directive. This guidance states that for non-hazardous but non-inert wastes a landfill lining system should consist of:

■ a leachate drainage layer of at least 0.5 m thick

■ an artificial sealing layer such as a geomembrane or dense asphaltic concrete (DAC)

■ a natural geological barrier (this may be artificially compacted to reduce its permeability) that ensures there is no discharge of 'List I' substances into the water table and no pollution of groundwater by 'List II' substances.

List I and List II substances are given in the Groundwater Directive (Council of the European Communities, 1979), which states that List I substances are to be eliminated and discharges of List II substances are to be reduced. List I and List II are included in Appendix 2.

The precise specification of the layers must be determined on a case-by-case basis by undertaking a risk assessment and modelling work. The risk assessment must take account of:

■ the operational phase of the landfill;

■ the subsequent period of active post-closure site management; and

■ the final phase when passive control measures are used.

The assessment must also consider the failure of the artificial sealing layer and/or the leachate drainage layer.

Flow through composite liners

Giroud et al. (1989) derived an equation for the flow through a geotextile liner that is in contact with a mineral layer immediately below the geotextile. They assumed that flow through the geotextile takes place through holes in the membrane (rather than by diffusion through the membrane) and that the permeability of the underlying mineral layer controlled the rate of leakage through the holes. The equation is

$$q = c \times a^{0.1} \times h^{0.9} K_s^{0.74}$$

where

q is the leachate flow per defect (m³ s⁻¹)

c is a constant depending on the contact between the membrane and the mineral layer (0.21 for good contact and 1.15 for poor contact)

h is the head of leachate (m)

a is the area of the holes (m²)

K_s is the permeability of the mineral layer (m s⁻¹).

Hall and Marshall (1991) suggested that a typical geomembrane will contain the following numbers of flaws per hectare (10 000 m²):

■ 5 pinholes (area 0.1–5 mm² per hole)

■ 2 small holes (area 5–100 mm² per hole)

■ 0.15 large holes (area 100–10 000 mm² per hole).

SAQ 25

(a) Estimate the annual flow of leachate through a composite liner consisting of a geomembrane above a 5 m thick mineral layer with the following characteristics and good contact between the geomembrane and mineral layers:

Site area	10 000 m²
Leachate head	0.75 m
Mineral permeability	1×10^{-9} m s⁻¹

(b) What would the leachate flow be if the geomembrane were to fail to such an extent that it could be considered to offer no protection at all?

(c) What would the steady-state leachate flow and depth of water be in the event of geomembrane failure and failure of the leachate pumping system if the effective rainfall infiltration is 40 mm y^{-1}?

(Hint: This will involve use of water balance and Darcy's law calculations.)

6.3.9   Leachate percolation and waste stabilisation

As rainwater flows through the landfilled waste, soluble components are dissolved and eventually all the soluble material will have been removed and the waste can be considered to be completely stable. This is important because it can be argued that, until the waste reaches this stage, it still has the potential for causing environmental pollution.

However, this can be an extremely long process. In a report on operating landfills as 'flushing bioreactors' the Chartered Institution of Wastes Management (1996) estimated that to reduce leachate pollutant concentrations to acceptable discharge levels, a 30 m deep landfill would require a flushing rate of 6 m per year for some 50 years. This should be compared with a typical UK rainfall of around 0.6 m per year, which suggests that landfills would need to be heavily irrigated to reach this condition in a reasonable timescale. Such a practice would contravene the Landfill Directive which aims to prevent the ingress of water to landfills.

In a similar calculation, Robinson and Latham (1993) showed that it would take 100 years for incoming rain to permeate to the base of a 30 m deep landfill.

6.3.10   Migration beyond the liner

The escape of leachate from the landfill site is only the first stage of the process that could, potentially, lead to the pollution of the wider environment. To assess fully the risk to the environment from landfill leachate, the following processes and pathways as illustrated in Figure 9 must be considered:

- migration through the artificial geological barrier
- migration through the natural mineral barrier taking account of the potential for the adsorption of pollutants onto the mineral surface, the oxidation of organic pollutants and the dilution of pollutants due to the mixing of the leachate with migrating surface waters
- migration of pollutants through the ground between the site barrier and the aquifer
- dispersion of pollutants in the aquifer.

This subject is considered in the final part of this section.

6.3.11   Leachate treatment

READ

File Article 3 in Appendix 1 describes a programme put in place to extract and treat the leachate from one of the UK's largest landfill sites. You should read this article now.

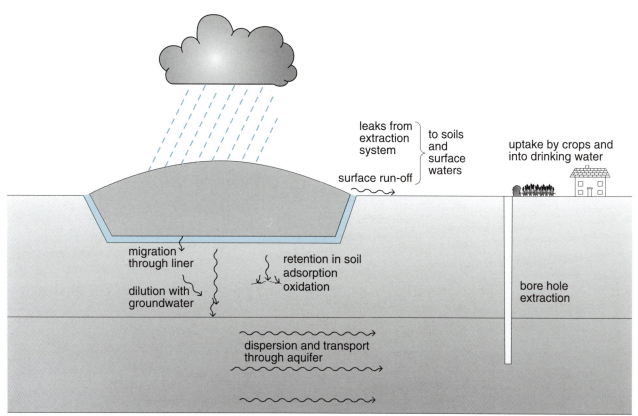

Figure 9   Leachate migration pathways

Under the Landfill Regulations (summarised in Section 6.2), all landfill sites are required to collect and treat leachate before discharging it to the environment.

The standards that must be reached before leachate can be discharged are determined by discharge consents. In deciding the conditions to be attached to these, the national regulator will be concerned with such factors as organic loading, and the presence of metals, sulphides, chlorinated hydrocarbons, ammonia, pesticides and herbicides.

In addition, the conditions attached will reflect:

- the type of waste deposited
- expected variations in flow rate
- expected variations in concentration of toxic compounds
- the nature and point of discharge.

Criteria will, therefore, generally be site-specific.

Treatment may be either on-site or off-site. Estimates produced by DEFRA (2004a) suggest that English landfills taking municipal and other biodegradable wastes produce about 12 million cubic metres of leachate per year and 2–6 million cubic metres per year can be attributed to MSW. DEFRA also estimated that 2% of this leachate is treated on-site and discharged to surface waters, 30% is treated on-site then discharged to sewer and 68% is discharged to sewer without any pretreatment.

On-site treatment can be achieved by:

- land irrigation, where leachate is sprayed onto land adjacent to the landfill site
- recirculation through the landfill site itself
- treatment in a specially built treatment plant.

Land irrigation

The use of land irrigation requires that a sufficiently large area of vegetated land is available to avoid overloading the natural processes that will be exploited to treat the leachate. When leachate is sprayed onto the land a significant reduction in leachate volume is achieved as a result of evaporation and transpiration. Evaporation can be maximised by using standard agricultural sprays.

The leachate then percolates through the soil, thus providing an opportunity for:

■ the microbial degradation of organic compounds

■ the removal of ions by precipitation or ion exchange

■ the possibility of the rapid uptake of constituents such as ammonia by the plants.

It is possible that spraying leachate will lead to the spreading of harmful pathogens: evidence to date, however, suggests that this is not a problem. Little information is available on the long-term effects of spraying leachate onto land. Regular monitoring needs to be carried out on both the leachate and the soil in order to ensure that metals and persistent organic compounds do not build up to unacceptable levels.

Recirculation through landfill

Recirculation of leachate through the landfill itself provides the opportunity both for reducing its strength and for reducing its volume by evaporation. At its simplest, recirculation involves collecting the leachate from the base of the site and spraying it onto the surface of the waste. Most landfills have practised leachate recirculation at some point during their lives. This is carried out to:

■ increase the rate of decomposition and reduce the time taken for the site to reach the methanogenic stage

■ reduce the very high BOD and COD of the leachate produced in the acetogenic phase

■ reduce the load on the treatment or storage plant during periods of high leachate generation.

However, it is possible that this practice will no longer be permitted under the IPPC regime.

Treatment in specially built plant

The treatment of leachate before discharge to sewers or to surface waters has become more common in the UK in recent years. One of the most effective treatment methods is known as the sequencing batch reactor (SBR) process. This is based on the activated sludge process that is covered in T210 Block 3: *Water pollution control*, but is operated in a batch mode. Leachate is fed into a reactor and then aerated to reduce the BOD and COD. At the same time nitrification takes place, which converts the ammonia into nitrites and then nitrates. After the aeration phase the leachate is allowed to settle and the liquor is drawn off from the top of the reactor leaving the solid matter behind. If necessary, the effluent can then be treated in a denitrification stage in an anoxic reactor to reduce the nitrates to nitrogen gas.

READ

File Article 4 in Appendix 1 presents a case study of the installation and operation of a treatment facility based on the SBR process with an ozonation stage and final reed bed. You should read this article now.

SAQ 26

After reading File Article 4 answer the following questions.

(a) Why was the reed bed considered necessary and what advantage does it have over other processes?

(b) Which treatment process would reduce the chloride concentration in the final effluent?

(c) Which pollutants would the ozonation stage treat?

(d) Why did the regulator feel justified in raising the discharge standards for chloride and COD?

One difficulty with leachate treatment is that the volume and strength of the leachate are subject to large seasonal variations. This can be minimised by balancing the flow, either in lagoons or by utilising the storage capacity available within the landfill site. The consequent reduction in peak loadings reduces the required treatment capacity.

Another potential problem is that the concentrations of the various compounds in the leachate change as the leachate ages. As discussed previously, leachates from newly deposited waste have a high concentration of readily biodegradable material that is amenable to biological treatment, whereas those from older sites are lower in organic content and are less easily biodegraded. For such 'stabilised' leachates the most common treatment objectives are the removal of ammonia, remaining BOD, COD and colour.

6.4   Leachate contamination risk assessment

6.4.1   Background

> **READ**
>
> Chapters 9 and 12 of T210 Block 1: *The environment, risk and public health* introduces the concepts of risk and risk assessment. You should re-read this material before you work through the following section.

In summary, an environmental hazard is something that has the potential to cause harm to health or to the environment, and a risk is the probability of the hazard occurring. So leachate is a hazard because it contains harmful substances. The risk from leachate depends on the likelihood of it escaping from the site and entering a watercourse, combined with the concentration of the hazardous components in the water and the use of the water.

So, a given concentration of, say, toluene in an aquifer will present a low (or even insignificant) risk if the aquifer is small and saline and there are no plans to abstract water from it for any purposes. By contrast, the same concentration of toluene could present a significant risk if the aquifer were used to abstract drinking water for a large conurbation.

This means that risk assessments need to be undertaken on a case-by-case basis. However, the regulations do set some boundaries. Under the Groundwater Regulations (HMSO, 1998) that implement the Groundwater Directive (Council of the European Communities, 1979) there must be no discharge to groundwater of List I substances and no pollution of groundwater by List II substances

(Appendix 2). As part of the IPPC application, site developers have to demonstrate that the site will meet these criteria. This is normally done by means of a risk assessment.

The process of risk assessment identifies potential sources of pollution (or hazard), receptors (people, other species or locations in the environment likely to be affected) and the pathway between the source and receptor. Unless all three elements are present there can be no risk. Even if a potential hazard exists, there is no risk if there is no receptor to be affected or no pathway between the pollutant and receptor.

6.4.2   The assessment process

Detailed guidance on undertaking risk assessments on the potential for pollution from leachate has been produced by the Scottish Environment Protection Agency (SEPA, 2002) and the Environment Agency (2003c). The risk assessment procedure is summarised in the following paragraphs.

Stage 1 – Characterisation of the location

The potential site needs to be clearly specified both in terms of grid references and also proximity to villages or towns, altitude, aspect and drainage. References should be given to the appropriate geological and hydrogeological maps.

The local geology then needs to be determined to identify the formations present, the depth of each layer and any fissures, fractures or other potential leachate transport pathways. This will involve the scrutiny of geological maps, other published material and the results of any field investigations carried out during previous uses of the land. In the absence of reliable data there may be a need to undertake additional field work involving borehole drilling.

Any aquifers present need to be identified and characterised in terms of permeability and importance. This will require reference to groundwater vulnerability maps and hydrogeological maps of the region.

The local groundwater must then be characterised and, unless recent local borehole data are available, this will necessitate the drilling and monitoring of boreholes to determine groundwater levels, flowrate, direction of flow and chemical composition (with particular reference to List I and II pollutants (see Appendix 2). SEPA (2002) recommends that monitoring should be carried out for at least 12 months before finalising the design to allow for seasonal fluctuations in flows.

Finally, any abstraction points and uses for the groundwater must be identified.

Stage 2 – Leachate characterisation

The amount of leachate likely to be produced should be predicted on the basis of water balance calculations as discussed in Section 6.3.7 above. Estimates of its composition will also be required and these will usually be based on data from existing landfills taking similar types of waste.

Stage 3 – Identification of compliance points (or receptors)

Identification of the aquifer abstraction points and uses of the abstracted water will help to determine the receptors. However, to comply with the Groundwater Regulations the compliance point (or receptor) for List I substances will be the point at which the leachate enters the aquifer (so the only attenuation processes will be those taking place in the zone above the aquifer). In practice, the compliance point is usually taken to be a borehole at the cell or site boundary.

For List II pollutants, the compliance point is allowed to take account of some dilution in the aquifer, but the compliance point will usually be a borehole adjacent to the site.

Stage 4 – Selection of environmental assessment limits

At present there are no standards for potential pollutants in groundwater, so environmental assessment limits (EALs) need to be determined to indicate the concentration of a particular pollutant that may impair the quality of the water. For List I substances, the EAL will normally be the Environment Agency's minimum reporting value (MRV) – in effect the minimum value that can be detected reliably. For List II substances, EALs are determined by considering other water quality standards such as those specified by the EU Drinking Water Directive, the UK Drinking Water Standards or the WHO Drinking Water Guidelines. There are cases where it would be inappropriate to select one of these standards for the EAL, for example:

■ When the natural level of a given substance in the aquifer is greater than the EAL it may be acceptable to set the EAL to the background level.

■ When the level of a given substance in the aquifer is greater than the standards due to pre-existing pollution, the EAL should be selected to a level that results in no further deterioration of the aquifer.

■ When the natural level of a given substance is substantially lower than the most stringent water quality standard, the EAL may be set at a point between the background level and the water quality standard.

The substances for which EALs are set should be determined on a case-by-case basis depending on the nature of the waste deposited and the likely composition of the leachate.

Examples of EALs for a number of sites are shown in Tables 32–4 (data from Environment Agency, 2002b).

Table 32   Example environmental assessment limits (EALs) for an inert waste landfill in clay overlying non-aquifer limestone

Determinand	Maximum groundwater concentration (from site measurements) (mg^{-1})	UK Drinking Water Standard (mg^{-1})	Selected EAL (mg^{-1})
Ammonium	<0.2	0.5	0.5
Chloride	78	250	250
Magnesium	23	50	50

Drinking water standards are selected as the EALs because it is not impossible that minor water abstractions may take place in the future.

Table 33   Example EALs for a non-inert waste landfill overlying a major aquifer

Determinand	Maximum groundwater concentration (from site measurements) mg l^{-1}	Typical leachate content (from the literature) mg l^{-1}	UK Drinking Water Standard mg l^{-1}	Environmental quality standard for fresh water mg l^{-1}	Selected EAL mg l^{-1}	Leachate: EAL ratio (AF_u)
Ammonium	<0.1	4000	0.5	–	0.5	8000
Chloride	28	8000	250	250	40	200
Toluene	<0.004	200	–	50	0.004	50 000

The EAL for toluene (List I) is taken to be the Environment Agency's MRV.

The EAL for chloride is taken as a value between that of an uncontaminated aquifer and the Drinking Water Standard value.

The high AF_u ratio for toluene (which means that attenuation and dilution processes will need to result in a reduction in excess of 50 000) suggests that it is highly unlikely that this will prove to be a suitable location.

Table 34   Example EALs for a non-inert waste landfill in clay overlying a minor aquifer

Determinand	Maximum groundwater concentration (from site measurements) mg l^{-1}	Typical leachate content (from the literature) mg l^{-1}	UK Drinking Water Standard mg l^{-1}	Environmental quality standard for fresh water mg l^{-1}	List I MRV mg l^{-1}	Selected EAL mg l^{-1}
Ammonium	1.2	2000	0.5			1.5
Chloride	300	4500	250	250		375
Potassium	8	540	10	–		10
Toluene	<0.004	80	–	0.05	0.004	0.004

For ammonium and chloride, where the groundwater exceeds the standards the EALs are based on the groundwater concentration plus 25%.

For potassium, the EAL is based on the Drinking Water Standard.

Stage 5 – Selection of trigger levels

Trigger levels are the levels of a given pollutant in the aquifer that would demonstrate that pollution has occurred as a result of the development. In setting trigger levels the main considerations are:

■ the substances for which trigger levels are set

■ the trigger level concentrations

■ the location of the monitoring points.

Generally, the trigger levels for List I substances would be the analytical detection limits and for List II substances they would be the EALs. Monitoring points are required for each potential pathway and receptor.

In addition to trigger levels, control levels are also to be set. If the routine groundwater monitoring programme detects pollutant levels at or above the control level, the regulator is to be informed and further assessments made. In the extreme case of the trigger levels being reached, a programme of remedial action must be agreed with the regulator.

Stage 6 – Prediction of groundwater contamination

This requires the modelling of the following stages as illustrated in Figure 9 above. I have considered the migration of leachate through the site liner in

Section 6.3.8 above, but to complete the picture the following must also be considered:

- vertical migration of the leachate through the unsaturated and saturated layers above the aquifer
- adsorption of metal and other species onto the ground materials during migration
- oxidation of organic species during migration
- the retardation of pollutant flows by adsorption and desorption processes
- horizontal migration of the leachate through the aquifer
- dilution, dispersion, adsorption and oxidation while travelling through the aquifer.

Models are available that carry out this assessment, but they are highly complex. The most widely used is the Landsim™ model developed on behalf of the Environment Agency. Landsim adopts a probability-based approach which means that the model is run a large number of times selecting different values for each parameter that may vary (for example: the range of likely rainfalls from a minimum to a maximum value; the number of defects in a synthetic liner). The simulation produces a very large number of outputs that are presented in terms of confidence levels. For simulations of this nature, a 95% confidence level will be set so the modeller can be 95% confident that the results presented will be better than the predicted levels.

Figures 10 and 11 are sample outputs produced from Landsim (Hall et al., 2003). Figure 10 shows the predicted leachate rate for a site with a 30-year operational period followed by 60 years post-closure leachate control. Beyond this time, leachate is no longer removed so the head of liquid rises, leading to increased leachate flows (as predicted by Darcy's law). In addition, the model takes account of the degradation of the liner and capping layer. After about 1000 years the site reaches steady conditions with a leachate flow of about 24 m^3 d^{-1}.

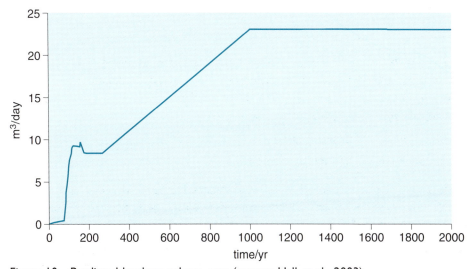

Figure 10   Predicted leachate release rate (source: Hall et al., 2003)

Figure 11 shows the predicted concentration of chloride in the receiving groundwater over a similar period. The initial rapid rise corresponds to the increase in leachate release when leachate removal ceases. The chloride concentration then decays and, after about 1400 years, all the soluble chloride has been leached out of the site.

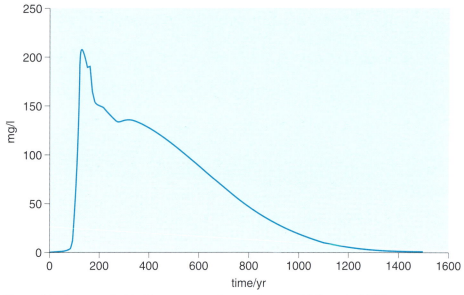

Figure 11   Impact of chlorides on groundwater (source: Hall et al., 2003)

SAQ 27

Considering the information in Tables 33 and 34, would you consider that a landfill site overlaying a major or a minor aquifer with the background chloride levels shown in these tables designed to give the releases shown in Figure 11 is environmentally acceptable?

This prediction stage of the assessment process must take into account the entire lifetime of the site including the post-closure phase. For these latter stages, factors such as the end of active leachate extraction and failure of any synthetic liner must be considered. The fact that the concentration of potential pollutants in the leachate may decline with time may also be taken into account.

Stage 7 – Assessment of results and consideration of mitigation methods

In the final stage of the assessment the results of any modelling should be compared with the EALs. Care should be taken to ensure that the results used are conservative, based on worst-case events and over the entire active life of the proposed site.

SAQ 28

What assumptions should you make when considering the worst case?

Consideration should then be given to measures that could be taken to reduce the impacts and to assessing the effectiveness of such measures.

SAQ 29

List the measures that could be taken to reduce the likelihood of harm to human health or to the environment from landfill leachate.

6.5   Landfill site development and operation

6.5.1   Site development

Before any waste can be deposited, the site infrastructure must be provided and basic site engineering works must be carried out.

The infrastructure will need to be carefully planned to ensure efficient and safe operation of the site throughout its life and includes provision for:

- the site entrance, which should be designed to ensure good visibility
- site accommodation, for example: messroom, garage, general store, fuel store, workshop, site control office, weighbridge, possible provision of offices for management and technical support staff, a small laboratory for analysis of incoming waste if the site is to take difficult industrial waste
- site notice boards
- wheel cleaning facilities to prevent mud being taken onto the highway from the site and causing a danger to traffic
- primary site road from the site entrance to the control office
- secondary roads from the weighbridge to the filling area
- security fences to prevent unauthorised site access
- litter fences to collect windblown litter and prevent it from leaving the fill area.

Necessary steps must also be taken to protect the equipment of undertakers such as electricity, gas, water, telephone, rail, mining and airports, if this is to be affected by the site development.

Basic site engineering works may include: earthworks, site lining, leachate management, monitoring, landfill gas management.

6.5.2   Earthworks

Examples of the functions of earthworks are in the:

(a) construction of the site roads and entrance;

(b) preparation of the cells to be used in the first phase of the operation;

(c) provision of earth banks to screen the operation from the surrounding area;

(d) construction of drainage ditches.

Extensive earthworks may also be necessary if artificial liners are to be used. During these operations many tonnes of earth may need to be moved and, where possible, this should be stockpiled for subsequent use.

6.5.3   Site lining

Site lining has already been discussed in Sections 6.2 and 6.3.8, so only the following points need be added.

Manufacturers suggest an expected life of 20–30 years for synthetic liners but these estimates are not based on actual in-service data. A summary of a series of laboratory tests is given in Table 35.

Table 35   Summary of findings from laboratory testing of synthetic liners

1	Liner material swelling and loss of strength was accelerated at high temperatures.
2	Not all liners within the same category had identical composition or physical properties as formulation can vary.
3	Liner compatibility with hazardous wastes is highly waste-specific. Oily wastes had the greatest effect on liner properties.
4	Leachate caused liners to swell and lose strength.
5	Chlorinated polyethylene, chlorosulphonated polyethylene and neoprene showed the greatest changes in physical properties. Low-density and high-density polyethylene performed well in all tests.
6	Seams were a weak point and several types of liner showed a loss in seam strength, particularly if adhesives were used. Heat-sealed seams retained their properties well.

Source: DoE (1997).

Installation of liners is highly specialised. Improper installation is the primary reason for failure of synthetic liners. The site needs to be carefully prepared beforehand. For example, if a smooth site base cannot be provided, it may need a blinding layer of fine-grained material to be installed below the site to prevent damage to the liner.

A similar protective layer is also installed above the liner to prevent puncturing by rocks or refuse. Drainage facilities for leachate collection can be incorporated in the protective layers. In the USA it is common practice, particularly for hazardous waste landfills, to install two synthetic liners with a permeable layer between them. The permeable layer can house sensors to identify any leachate percolation. Monitoring at several sites has shown that leachate does penetrate the upper liner. File Article 5 describes such a two-layer system installed at a site in Germany.

> **READ**
>
> Read File Article 5 in Appendix 1 now.

6.5.4   Leachate management

Surface water will need protecting. It may be necessary to divert existing watercourses as well as to deal with water running off adjacent land and surface water from the site itself. Surface run-off can frequently be intercepted with a peripheral ditch system.

Groundwater should be prevented from entering the site. To do this it may be necessary to excavate down to impervious strata and seal the site with clay or bentonite backfill, or to provide an artificial interception and drainage system.

Leachate will need to be collected for treatment. At small sites, the base of the site can be graded to a low point or sump so that leachate can be collected and pumped away.

As mentioned above, the use of a permeable material as the protective cover for the liner will also provide a flow path for the leachate. In larger sites a perforated drainage pipe system can be used. If the leachate is to be treated before removal from site, tanks or lagoons will need to be provided, together with a system of pipework and pumps for transfer.

6.5.5   Monitoring

It is essential to install a monitoring system to confirm that groundwater is not being contaminated. Monitoring boreholes will need to be provided around the site, and the number, design and location of these will have been agreed with the regulator at the planning stage.

It is common to use stacked perforated rings to construct the monitoring wells. Further rings can be added as filling progresses. This system has the added advantage that a suitable submersible pump can be used in the wells if leachate has to be removed.

6.5.6   Landfill gas management

Landfill gas management techniques discussed in Section 6.2.7 will need to be implemented as appropriate.

6.6   Site operation

The main features of the site operation will have been set out in the working plan developed previously. The planning of operations is influenced by:

- the type and quantity of wastes to be landfilled
- whether or not the wastes are to be pretreated
- the need to maximise site life
- the type of delivery vehicles to be used
- the expected current and future rates of filling
- the type and availability of cover material.

Virtually all UK landfill sites operate using the *cell method,* in which waste is deposited in preconstructed *bunded* areas. This method is becoming widely used and is the preferred method of operation in the UK, since it encourages progressive filling and restoration.

Figure 12 shows a typical operational plan for a landfill site using the cell method.

Leachate generation can be minimised by paying careful attention to cell design. A number of factors have to be considered in deciding the optimum size of each cell:

- quantity and variation of rainfall
- absorptive capacity of the waste
- rate of waste input
- number of incoming vehicles
- the need to ensure sufficient working space for safe vehicle turnround.

In some sites, daily cells may be constructed within the larger main cells.

The cell walls are constructed either by pushing material up from the base of the site (on the initial lift) or from waste material. In either case, care needs to be taken to ensure their structural stability. In addition to concealing the operation, the cell walls help reduce the incidence of windblown litter.

The working face should be of a size that allows safe manoeuvring of vehicles and site plant, yet minimises infiltration, cover requirements and litter problems. The size of the working face needs to be reviewed regularly to ensure that it is at its optimum.

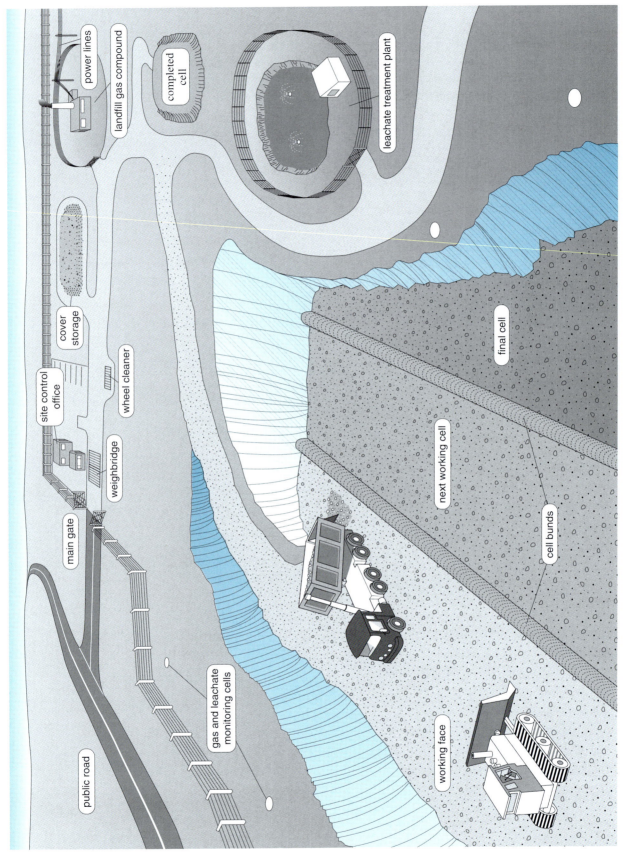

Figure 12   Landfill operation – the cell method

Incoming waste should be placed in such a way as to achieve a high degree of compaction. WMP 26 (DoE, 1997) lists the following advantages of compaction:

(a) an increase in waste density leads to an extension of the life of the site;

(b) a uniform, well-compacted layer of waste reduces the volume of daily cover required;

(c) a well-compacted site is visually more acceptable and carries less risk of litter blowing across the site;

(d) compaction reduces the incidence of fly infestations and colonisation by vermin;

(e) voids are eliminated, thus largely preventing underground fires, while surface fires become much easier to control;

(f) well-compacted waste provides a more stable base for delivery vehicles during discharge of loads. This reduces vehicle wear and tear and the risk of machinery becoming bogged down during wet weather;

(g) a high degree of compaction reduces the degree of settlement while ensuring that it takes place more evenly.

Good compaction is helped by placing the waste in thin layers (approximately 0.3 m thick) and running over each layer several times, using a steel-wheeled compactor. Such compactors are increasingly used on landfill sites. Initial densities of more than 1 t m^{-3} are achievable using the above method – considerably higher than the 0.5 t m^{-3} that can be achieved with a tracked vehicle pushing waste from the top of the working face. At the end of each working day all faces should be covered to a depth of not less than 0.15 m using a suitable covering material. Daily cover is essential for good site operation, since it reduces odours, inhibits pests and flies, helps control infiltration, minimises the risk of fires and improves the general appearance of the site.

One disadvantage of the cell method is that it can use up a lot of void space. This can, however, be overcome by using one of two strategies:

1   making the cell walls from suitable incoming waste;

2   selectively removing walls at the end of each day for use as cover.

When the final level is reached, the cell is capped to reduce leachate generation.

Throughout its operational life, the site should be monitored to ensure that should difficulties arise, the necessary remedial action can be taken.

6.6.1   Restoration and after-use

Restoration is site-specific and the restoration plan will have been produced as part of the working plan. It is important that restoration is considered at the outset for any landfill project. It is a requirement of the Landfill Directive that the prices charged by an operator reflect the cost of site restoration and aftercare for a period of at least 30 years after closure of the site.

The after-use of a site should also be considered at the early planning stages, in order to ensure that the best environmental and amenity benefit can be obtained from the project. Generally, land is returned to agricultural use (for grazing or arable crops), forestry or general amenity use. There is, however, scope for a more imaginative approach, particularly where the site is close to urban areas. Other options that may be considered include: sports grounds, golf courses, open storage, vehicle parking.

Building on completed landfill sites is not recommended until the site has become stabilised. The time required for this will depend on the waste types, the degree of compaction achieved and the moisture content of the fill.

The capping requirements specified in the Landfill Directive have already been summarised in Table 24, but the key features are discussed in the following paragraphs.

The installation of a cap of low permeability (less than 1×10^{-9} m s^{-1}) helps to reduce the quantity of leachate generated. The cap should be either domed or contoured to encourage surface water run-off. Clay is the most frequently used material for capping landfill sites, although other materials used for lining sites are possible substitutes. Synthetic materials have the disadvantage, however, that they are easily damaged and have no self-sealing properties.

A cap thickness of 1 metre for natural materials has been found to be effective. The amount of settlement must be anticipated in achieving the agreed final contours, and an even settlement can be promoted by good site operation during the operational phase. Uneven settlement can be a major cause of cap failure. Most of the settlement will have taken place within 10 years of site completion, although final levels may not be reached for up to 30 years.

The cap should be covered as soon as possible with soil and vegetation, to help to prevent water infiltration and to encourage water loss by transpiration. The choice of vegetation is important, since the integrity of the cap must not be endangered by deep roots which can penetrate it. Most arable crops require a 1-metre depth of soil and although a reduced thickness can be tolerated for grassland and recreational after-use, 1 metre is generally needed to protect the cap from damage caused by machinery, drying and cracking, plant root penetration, burrowing animals and erosion.

The quality of the soil protecting the cap is particularly important if it is to be cultivated. Good quality topsoil, however, is expensive to buy. Soil stripped at the beginning of the operation can be stored in screen bunds for use in restoration. Good quality soil coming in during the operation should if possible also be stockpiled. The ability of the operator to do this may be limited by site constraints, although the modern practice of progressive restoration will help alleviate the problem.

Where soil quality is poor, it can be improved by incorporating bulky organic materials. Composted waste is often used for this purpose, but other materials such as conditioned or digested sewage sludge, milk, whey, brewery waste and spent hops, spent mushroom composts and sugar beet wastes have all been used.

However, in using such materials care must be taken to ensure that excessive quantities of contaminants such as heavy metals from sewage sludge are not introduced. Soils may also need to be improved by liming and by fertiliser addition.

The chosen final landform must be designed to blend in with the surrounding land as well as to promote surface water run-off and drainage. The gradient is of particular significance. A minimum of 1 in 30 is usually considered adequate to prevent ponding and drainage problems caused by differential settlement. Steep gradients should be avoided, particularly if the land is to be used for arable crops, where the maximum gradient should be less than 1 in 6. Drainage can be assisted by installing underground drainage systems.

The problems associated with building on completed landfill sites include differential settlement and effects on structures and foundations caused by gas and leachate.

Some building on old sites has taken place, but its success is difficult to judge because of the short time that has elapsed since the construction. Engineered containment sites should not be built on because of the possibility of compromising the containment by destroying the integrity of the cap.

For other sites a full site investigation is necessary to assess the potential problems. Factors to be considered include:

- assessment of the underlying strata
- performance of the waste as a foundation material
- possible chemical attack
- gas and leachate generation
- toxicity
- nature of fill (amount of biodegradable material, homogeneity, chemical composition, presence of combustible material)
- degree of compaction
- depth of fill
- landfill age
- likely settlement (general and differential).

It should be emphasised that restoration cannot be considered to be complete even when the soils are emplaced and the land has been engineered to the design contours. The land will need to be carefully managed for a number of years to help the soil recover from the effects of movement, storage and replacement. If the site is to be used for agriculture, forestry or amenity, there will be a need to prepare and carry out a management scheme to bring the site to the required standard and to maintain it. In addition, works to remedy differential settlement will need to be carried out, and leachate and gas will need to be monitored and controlled.

It is important to stress that a site operator remains responsible for the site until it can satisfy the regulator that the site poses no threat to the environment. The wording of the Landfill Directive suggests that this will take 30–50 years after site closure. However, sites may need active management for a century or even longer.

6.7   Health impacts of landfills

6.7.1   Public health effects

The DEFRA review that I considered in Section 5.2 also considered the impact of landfills. Like incineration, much of the public concern over landfills stems from the emissions from poorly controlled sites in the time before the tightening of regulations.

DEFRA identified the key study in this area to be the one carried out by the Small Area Health Statistics Unit (SAHSU) (Elliot et al., 2001). This study looked at birth defects and low birth weights in the population living in the zone within 2 km of landfill sites operational at some point between 1982 and 1997. The control group consisted of people living more than 2 km from a landfill site. This study covered 774 hazardous waste sites (many would have been co-disposal sites accepting only a small percentage of hazardous waste),

7803 non-hazardous waste sites and 998 sites accepting waste of unknown content. It is interesting to note that the 'exposed' group living within 2 km of a landfill accounted for 55% of the UK population.

Elliot et al. examined the data on 124 597 congenital abnormalities (including terminations) amongst the 8.2 million live births and 43 471 still births in the study and control groups. After controlling for deprivation, they found a small but statistically significant increase in abnormalities and low birth weights as shown in Table 36.

Table 36   Increase in birth defects for populations within 2 km of a landfill

Outcome	Usual rate	Observed increase	99% confidence interval
Neural tube defects (1)	1 in 1800 births	6%	1–12%
Cardiovascular defects	1 in 750 births	No increase	
Hypospadias and epispadias (2)	1 in 420 births	7%	4–11%
Abdominal wall defects	1 in 2900 births	7%	1–12%
Gastroschisis (3) and exomphalos (4) (also included in the above category)	1 in 5300 births	18%	3–34%
Still births	1 in 195 births	No excess	
Low birth weight (<2.5 kg)	1 in 16 births	6%	5.2–6.2%
Very low birth weight (<1.5 kg)	1 in 104 births	4%	3–6%

Source: Data from DEFRA (2004b).

1. Anencephaly (partial or complete absence of the baby's brain) and spina bifida (an opening in the spine requiring surgery soon after birth to close the spine and prevent further damage).

2. The urethra does not develop fully.

3. A herniation or displacement of the intestines through a defect on one side of the umbilical cord.

4. Herniation of the intestine into the base of the umbilical cord.

DEFRA pointed out that, whilst Elliot et al.'s work represents the strongest piece of epidemiological research on the health effects of landfills, it still suffers from some deficiencies. Most importantly, it was not able to state that the increases in birth defects were due to emissions from the landfills or due to some other cause or combination of causes. DEFRA also noted other possible weaknesses of the study:

■ small errors in adjusting for socio-economic factors could have an appreciable impact on the results;

■ the study was based on residency at the time of birth which was not necessarily the residence during the critical early stages of pregnancy.

In addition, some of the sites studied opened during the course of the programme and the residents experienced a reduction in birth defects after the opening of the site.

The SAHSU study also looked at the incidence of a number of adult and childhood cancers (Jarup et al., 2002). Due to the long latency period in developing cancer, the results are less reliable than those for birth defects. However, after controlling for socio-economic factors, they found no excess of cancers among the population living within 2 km of a landfill site.

Like incineration, landfill accounts for a relatively small proportion of the UK's pollutant inventory, as shown in Table 37. However, it is a major source of methane (second only to agriculture) and, perhaps more surprisingly, of cadmium. The methane is due to the escape of landfill gas through site liners and caps, and discharges from old sites without active gas extraction and combustion plant. The cadmium is emitted in the landfill gas (and its combustion products) and is a direct consequence of cadmium compounds in the raw waste.

Table 37   Contribution of landfill to UK pollutant emissions

Pollutant	Contribution of MSW landfill to UK total (%)
Carbon dioxide	0.8
Methane	28.6
Particulate matter	0.06
NO_x	0.4
SO_2	0.16
HCl	0.31
Dioxins and furans	0.53
Cadmium	9.8
Mercury	0.10

Source: Data from DEFRA (2004b).

6.7.2   Health impacts on site operators

Landfill site operators are potentially exposed through inhalation to dust, bioaerosols and volatile organic compounds (VOCs). Consequently the Health and Safety Laboratory (Swann et al., 2004) undertook a monitoring programme at five UK landfills to investigate worker exposure.

The main findings of this study were:

■ VOC concentrations were higher than ambient levels, but much lower than the relevant occupational exposure limits. Therefore they were not considered to be a significant risk.

■ Bioaerosol levels varied with sampling location; bacterial levels of up to 10^6 cfu m^{-3} and fungal levels of 10^4 cfu m^{-3} were measured although worker exposure was less.

■ Worker exposure to bioaerosols was similar to those experienced in other waste management and agricultural activities.

■ Vehicle cabs were found to be effective in reducing worker exposure, but this protection was reduced if workers had to leave the cab or chose to keep the windows or doors open while working.

7 NOVEL WASTE MANAGEMENT PROCESSES

7.1 Background

So far, this block has focused on the conventional technologies that are currently used to treat and dispose of virtually 100% of the UK's wastes. However, there are other systems that are still to be proven at the commercial scale.

Policy and regulatory measures such as the implementation of the Landfill Directive and statutory recycling targets mean that these novel technologies are likely to become important in the future. However, the adoption of any new technology is trapped in a vicious circle. Only commercially proven technologies attract firm waste contracts and funding, and without contracts and funding commercial-scale plants will never be built.

The development of these technologies is changing rapidly and the Course Team will use the T308 website to keep you informed of developments in this area.

The Department for Environment, Food and Rural Affairs (DEFRA) is working with local authorities and the waste industry to break this circle. Under the Waste Implementation Programme (WIP), DEFRA has committed some £30 million to help fund demonstration projects that use novel technologies to treat and divert biodegradable municipal waste from landfill. A number of projects were identified in September 2004 based on MBP, gasification, pyrolysis and anaerobic digestion technologies. However, contractual and legal problems meant that progress has been much slower than hoped. At the time of writing (summer 2006) contracts had been agreed on the following projects:

- The 'Bioganix' in-vessel composting plant – producing compost from source-segregated kitchen waste.

- The 'Greenfinch' anaerobic digestor – operating at 25% of its 5000 t per year design capacity processing source-segregated kitchen waste.

- The 'Premier Waste' aerobic treatment plant – under construction in County Durham.

- The 'Envar' in-vessel composting system – under construction in Cambridgeshire. When operational, 105 000 t per year will be processed generating compost and refuse-derived fuel products.

- A gasification plant to be based at Dagenham, East London. This will take 90 000 t per year of RDF produced by a waste management company and produce 8–10 MW of power. However, the project has still to obtain planning permission.

7.2 The technologies

> **READ**
>
> Advanced thermal processing (gasification/pyrolysis) and anaerobic digestion are described in Section 9.4 and Chapter 10 respectively of T210 Block 4: *Wastes management*. You should reread this material now. The concept and technologies behind mechanical biological treatment are described below.

7.3   Mechanical biological pretreatment (MBP)

Mechanical biological pretreatment (MBP) is not in itself a method of disposing of waste, but it is a way of treating waste using a combination of mechanical processes such as shredding and screening, and biological processes such as composting or digestion. The main aim of MBP plants is to reduce the amount of unprocessed biodegradable waste being sent to landfill. The Landfill Directive was a strong driver in stimulating interest in these systems. Some materials that can be recycled are recovered during the process and fuel can also be recovered from the waste. The remainder is landfilled, but this stabilised organic fraction has a lower potential for causing environmental damage than the raw waste.

MBP is also known by a number of other names including mechanical biological treatment (MBT). However, I have used MBP to stress that the technology is only one stage in the treatment/disposal of waste and is not a solution to the waste problem in its own right. Several local authorities are considering using the technology and some significant installations are at an advanced stage of development. MBP is much more common in other European countries with 40 plants operating in Italy, 35 in Germany and 11 in Austria.

MBP is based on the technology developed in the late 1970s and early 1980s to manufacture a refuse-derived fuel (RDF) for use as a coal substitute in industrial boiler plant. Indeed some MBP plants are still designed to produce a fuel as one of their outputs. There are many designs of MBP plant available depending on the intended outputs from the plant, but the idea behind the process can be seen from the typical system shown in Figure 13 and described below.

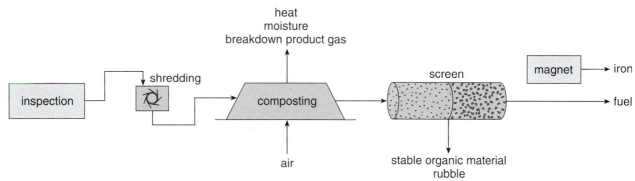

Figure 13   Diagram of an MBP plant

1   The incoming waste is tipped from the vehicles in an enclosed building and inspected so that any large items that might damage the equipment can be removed.

2   The waste is then shredded to reduce the size of the largest particles to 600 mm.

3   The shredded waste is placed in long rows in a building and air is passed through the rows for about 12–15 days. During this composting process the temperature reaches 50–60 °C, moisture is driven off by the heat and the readily degradable organic wastes break down. As a result, about 25% of the weight of material is lost. The remaining waste is more amenable to physical sorting, more stable and has a higher calorific value.

4   After processing, a series of screens, weight separators and magnetic separators is used to produce four products:

■   a paper and plastic-rich material for fuel use (RDF) although it may be possible to recover some of this material for recycling

■   metals for recycling

■   a mixture of glass, stones and rubble for use in landfill construction

■ a stabilised organic fraction that is either landfilled directly or possibly processed further to reduce its biodegradability and then landfilled or used for landfill engineering, landscape or agricultural use.

Other MBP plants are less complex and only carry out the first three stages and are simply intended to reduce the weight of waste sent to landfill and its potential for producing gas and leachate.

7.3.1   Questions over MBP

At the time of writing, both the technology and regulation of MBP are still undergoing development. During the lifetime of this course it is expected that a number of plants will come on stream and these questions will be answered. The course team will use the T308 website to update you on these issues, but you should be aware that the following will need to be resolved.

Use of the fuel

One of the factors that stopped the development of RDF in the 1980s was the decision that RDF combustion was, in fact, a waste incineration process. There is no reason why this view will change so any plant that burns the MBP fuel will be classed as an incinerator, will be regulated under IPPC and will have to meet the requirements of the Waste Incineration Directive. Local authorities who adopt MBP to 'avoid the need for incineration' may find themselves having to support and grant planning permission for an incineration process after all. The alternative would be to export the fuel to a neighbouring authority, but for many people, this approach violates the 'proximity principle' and would be politically unacceptable.

Use of the biostabilised material

Biostabilised material cannot be classed as 'compost' and is not eligible for consideration under PAS 100 (see Section 4.2.4) because it is not derived from source-segregated waste. At present biostabilised material is regarded as a waste, which means it could only be applied to land under a waste management licence (such as a landfill) or as an 'exempt activity'. Exempt activities include using the waste to improve the quality of agricultural land, improve the ecology of land or as part of a process to reclaim land previously used for industrial purposes. Exempt activities do not need a waste management licence, but they still need to be registered with the national regulator and are tightly controlled.

Reduction in biodegradability

A key driver in the installation of an MBP plant is the reduction in biodegradable waste sent to landfill. This can be achieved in a number of ways. Firstly, biodegradable material such as paper may be screened from the waste stream and included in the fraction to be incinerated. Secondly, biodegradable material may be biologically stabilised and used beneficially in non-landfill applications. Thirdly, the biodegradable content of the waste can be reduced by composting or anaerobic digestion. These processes reduce the mass of organic material for disposal, but can also reduce the biodegradability and may contribute towards the Landfill Directive targets.

At the moment there is no standard test to determine the biodegradability of a waste sample. The scientific community is developing tests which may be based on the oxygen uptake and CO_2 release of a sample, loss in mass under aerobic conditions, loss of mass under anaerobic conditions or any combination of these. Once a test (or suite of tests) has been agreed, the regulations then need to be put in place. Until then, it is impossible to state to what extent MBP will reduce the biodegradability of a waste with respect to Landfill Directive diversion targets.

8 ENVIRONMENTAL AND FINANCIAL IMPACT OF WASTE

8.1 Assessing overall environmental impacts

8.1.1 Revision of BPEO, IPPC, BATNEEC, EIA

The concepts of best practicable environmental option (BPEO), integrated pollution prevention and control (IPPC) and best available techniques not entailing excessive costs (BATNEEC) are discussed in both the Environmental and Food Legislation Supplement and the T210 set book, *Dictionary of Environmental Science and Technology.*

Environmental impact assessment is discussed in detail in T210 Block 7: *Environmental impact assessment.* If necessary, you should read the relevant sections of the above material now.

In the following two sections I will examine two commonly used tools to determine and evaluate the environmental impact of a waste management strategy (or indeed of any other development). These tools are known as life cycle assessment and multi-criteria decision analysis. It should be stressed that they are merely aids to comparing different strategies; they will not take the decision for you.

8.1.2 Life cycle assessment

Introduction

Life cycle assessment (LCA) is a way of determining the impact of a process or product over its entire life (sometimes referred to as 'cradle to grave'). It looks at the environmental impacts (often referred to as 'burdens' in the LCA literature) of raw material extraction, processing and manufacturing, distribution use and disposal at the end of its life. The technique is often used to compare different brands of goods (for example detergents) or different ways of achieving the same ends (for example disposable and cloth nappies, petrol or diesel fuel). In recent years the technique has been applied to comparing different waste management options and strategies, and the Environment Agency has played a leading part in this in England and Wales.

Formal definitions of LCA differ in detail, and you may notice from the following three that each clarifies certain points. The Society of Environmental Toxicology and Chemistry (SETAC) originally defined LCA as:

> An objective process to evaluate the environmental burdens associated with a product, process or activity by identifying and quantifying energy and materials used and wastes released to the environment, and to evaluate and implement opportunities to affect environmental improvements. The assessment includes the entire life cycle of the product, process or activity, encompassing extraction and processing raw materials, manufacturing, transportation and distribution, use/reuse/maintenance, recycling and final disposal.

> (Fava et al., 1991)

A second definition is:

> Environmental LCA or product life analysis (PLA) are detailed studies of the energy requirements, raw material usage and water, air, and solid wastes generation of an activity, material, product or package throughout its entire lifecycle. Included are raw material sourcing, manufacturing/processing,

distribution, use/reuse/maintenance, and post-consumer disposal. The system under study is integrated in an input/output analysis and its material (raw material usage and waste generation) and energy flow are quantified. Based on this inventory, a final comparison among alternatives should be made, identifying opportunities for reducing energy requirements, raw material usage, emissions to water and air, solid waste generation and conserving natural resources (e.g. as part of a broader environmental auditing).

(de Smet, 1990)

A more recent description from the introduction to British Standard BS EN ISO 14040 (1997) is:

LCA is a technique for assessing the environmental aspects and potential impacts associated with a product by

- compiling an inventory of relevant inputs and outputs of a product system;

- evaluating the potential environmental impacts associated with those inputs and outputs;

- interpreting the results of the inventory analysis and impact assessment phases in relation to the objectives of the study.

LCA studies the environmental aspects and potential impacts throughout a product's life (i.e. cradle to grave) from raw material acquisition through production, use and disposal. The general categories of environmental impacts needing consideration include resource use, human health, and ecological consequences.

The description also notes: 'An inventory may include environmental aspects which are not directly related to the inputs and outputs of the system.'

SAQ 30

Notice that there are subtle differences in the definitions, but compare them and prepare your own summary of what LCA is all about.

The main use of LCA is to identify environmental improvement opportunities and potentially it is a powerful tool which:

- can assist regulators formulate legislation

- has the potential to identify opportunities to improve the environmental aspects of product and service systems at various points in their life cycle

- can help manufacturers to assess and improve their products and processes by, for example, identifying those parts of the process which have the potential to cause most pollution, or have a particularly heavy material or energy demand

- can assist in identifying where scarce resources are being used and may help in identifying areas where materials substitution can occur

- can help consumers make more informed choices

- can assist in the development of the various 'eco-labelling' schemes which are being planned or operated

- can help to choose between different products

- can improve the environmental performance of products or services

- can be used to support environmental labelling claims

- can be a valuable analytical tool to use in the decision-making process in conjunction with a number of important factors other than those addressed by LCA.

Methodology

One of the major findings of the 1990 SETAC workshop reported in Fava et al., 1991 was the consensus that complete life cycle assessments should be composed of three separate but interrelated components. It was recognised that existing LCA efforts had focused primarily on the inventory component.

The three components they identified are:

- Life cycle Inventory. An objective data-based process of quantifying energy and raw material requirements, air emissions, waterborne effluents, solid waste, and other environmental releases throughout the life cycle of a product, process or activity.

- Life cycle Impact Analysis. A technical, quantitative, and/or qualitative process to characterise and assess the effects of the environmental loadings identified in the inventory component. The assessment should address both ecological and human health considerations, as well as other factors such as habitat modification and noise pollution.

- Life cycle Improvement Analysis. A systematic evaluation of the needs and opportunities to reduce the environmental burden, associated energy and raw materials use, and environmental releases throughout the whole life cycle of the product, process or activity. This analysis may include both quantitative and qualitative measures of improvements, such as changes in product, process and activity design; raw materials use; industrial processing; consumer use; and waste management.

It was recognised that the LCA process is not necessarily a linear step-wise process and that environmental benefits can be realised at each step. For example, the inventory alone may be used to identify opportunities for reducing environmental releases, energy and material use. BS EN ISO 14040 (ISO, 1997) gives a broadly similar, if more detailed, breakdown of the constituent parts of the LCA process:

- definition of goal and scope
- life cycle inventory analysis
- life cycle impact assessment:
  classification
  characterisation
  weighting
- life cycle interpretation.

Each of these is described below. Table 38 lists the current status of ISO standards relevant to specific LCA elements or phases (as at January 2005).

Table 38   ISO standards

Standard	Title
BS EN ISO 14040: 1997	Environmental management – Life cycle assessment Principles and framework
BS EN ISO 14041: 1998	Environmental management – Life cycle assessment Goal and scope definition and inventory analysis
BS EN ISO 14042: 2000	Environmental management – Life cycle assessment Life cycle impact assessment
BS EN ISO 14043: 2000	Environmental management – Life cycle assessment Life cycle interpretation

Definition of goal and scope

A clear understanding of the goal of the study, the system(s) to be studied and the relevant requirements for critical review and communication of results are required at an early stage of an LCA. The goal, scope and intended application influence the direction and depth of the study, as well as the make-up and type of critical review effort.

The goal of the LCA should be defined and should include a clear and unambiguous statement of the:

- reason for carrying out the LCA
- intended use of the results
- intended audience
- initial data quality goals.

The scope of the study should define the:

- system
- system boundaries
- data requirements
- assumptions
- limitations (when known)
- type of critical review process to be employed
- type and format of the report required for the study.

It should be defined well enough to ensure that:

- the breadth and depth of analysis are compatible with, and sufficient to address, the stated goal
- all boundaries, methodology, data categories and assumptions are clearly stated, comprehensible and visible.

The SETAC conference concluded that the following should be the minimum study components to be defined;

- the product, processes or activity to be studied
- the reasons for conducting the study, including the needs and potential applications of pertinent user groups
- how the results of the study will be used by the group performing or sponsoring the study
- the elements of the analysis, for example: 'this life cycle assessment will determine and document the raw material and energy demands of the product being analysed, including all products or processes contributing to its manufacture and use, and the generation of wastes and co-products'
- the elements not addressed, for example: 'this study does not address socio-economic and aesthetic issues'.

Life cycle inventory analysis

The specific objectives for conducting the inventory include, but are not restricted to, the following (Fava et al., 1991):

- to establish a comprehensive baseline of information on a system's overall resource requirements, energy consumption and emission loadings for further analysis
- to identify points within the life cycle as a whole, or within a given process, where the greatest reduction in resource requirements and emissions might be achieved

■ to compare the system inputs and outputs associated with the alternative products, processes or activities

■ to help guide the development of new products, processes or activities toward a net reduction in resource requirements and emissions

■ to help identify needs for the life cycle impact analysis

■ to provide the information needed to conduct an improvement analysis.

Participants at the 1990 SETAC conference developed a technical framework for the key phases of a life cycle inventory. This is shown in Figure 14.

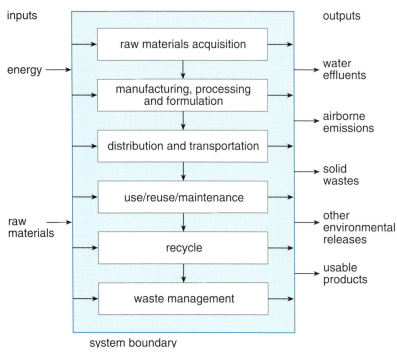

Figure 14   Life cycle inventory (source: Fava et al., 1991)

In general, each stage receives *inputs* of materials and energy and produces *outputs* of materials or energy which move to subsequent phases, and wastes that are released into the environment.

To be a truly useful technical tool a life cycle inventory should meet certain criteria. It should be:

■ scientifically based

■ quantitative

■ in appropriate detail

■ replicable

■ comprehensive

■ broadly applicable to the range of situations expected

■ consistent

■ peer reviewed

■ useful.

The process of undertaking an inventory analysis is iterative and the sequence of events will involve checking procedures to ensure that the requirements laid down in the goal and scope definition have been met.

Data should be collected for each element within the system boundary. The procedures used for collection will vary. Data collection can be a resource intensive process and there may be practical constraints in the data which can be collected.

Allocation procedures are an important consideration. Such issues are encountered when dealing with multiple inputs or outputs (e.g. generation of co-products, waste treatment processes and recycling). Specific rules should be used in allocating the materials and energy used in different inputs and outputs.

The presentation of the analysis should provide all the necessary elements to allow a thorough examination of the results. These elements comprise:

- the methodological options chosen
- information on the sites; sources of data; and models used to simulate the system studied.

Conclusions drawn must be consistent with the methodological choices made and the quality of data in the study.

Impact assessment

Analysis is only one part of the process, however. The data are helpful and informative but we need to be able to examine the findings and assess their meaning – assessment involves interpretation of the data and introduces 'value judgements' into the process. For example:

- Is the emission of x tonnes of one material more or less important in environmental terms than the emission of y tonnes of another?
- How can you compare a process which has a heavy energy demand with one that has a heavy material use or water demand?
- How do you compare the impact of landfilling waste, with its potential for pollution of water, emission of methane and possible general environmental loss of amenity, with incineration involving landfill of the ash and possible air pollution and water pollution?

Impact assessment is a technical quantitative and qualitative process which attempts to classify, characterise and value the magnitude and significance of environmental impacts, based on the information from the inventory analysis. There is increasing interest in this area and methods for impact assessment are still being developed. It is an iterative procedure and three steps are identified in BS EN ISO 14040 (1997), i.e. classification, characterisation and weighting.

Classification is the step in which the inventory parameters are grouped together and sorted into a number of impact categories.

Characterisation is the step in which the analysis and quantification of the risk of each of the impact categories takes place. There are various approaches for characterisation, e.g. relating data to environmental standards or modelling both exposures and effects and applying these on a regional or local basis. Each approach involves a different level of uncertainty, for example with determining the magnitude of each impact.

Weighting is the step in which the results are aggregated. This should be done only in very specific cases and only when it is meaningful.

Life cycle interpretation

This is the phase of the LCA when the findings of previous steps are combined to reach conclusions and recommendations. This process may be iterative and may include reviewing and revising the scope of the LCA as well as the nature and quality of the data collected. The findings should take account of any sensitivity analysis which may have been carried out.

Reporting

In general, LCA results should not be reduced to a simple overall conclusion, since trade-offs and complexities exist for the elements analysed at different stages of their life cycle. If an overall conclusion is made, the process used for valuation should be explicitly stated.

The results should be fairly and accurately reported to the intended audience. The report should cover:

- the goal(s) of the assessment
- the methods employed
- the results in adequate detail
- critical elements of the results (e.g. those identified by a sensitivity analysis).

The results, data, methods and assumptions should be presented in sufficient detail to allow the reader to understand interpret and use the results in a manner consistent with the goals of the study.

What problems are there with LCA?

BS EN ISO 14040 (1997) states:

> This International Standard recognizes that LCA is still at an early state of development. Some phases of the LCA techniques, such as impact assessment, are still in relative infancy. Considerable work remains to be done and practical experience gained in order to further develop the level of LCA practice. Therefore, it is important that the results of LCA be interpreted and applied appropriately.

> If LCA is to be successful in supporting environmental understanding of products, it is essential that LCA maintains its technical credibility while providing flexibility, practicality and cost effectiveness of application. This is particularly true if LCA is to be applied within small- and medium-sized enterprises.

> The scope, boundaries and level of detail of an LCA study depend on the subject and intended use of the study. The depth and breadth of LCA studies may differ considerably depending on the goal of a particular LCA study. However, in all cases, the principles and framework established in this International Standard should be followed.

> LCA is one of several environmental management techniques (e.g. risk assessment, environmental performance evaluation, environmental auditing, and environmental impact assessment) and may not be the most appropriate technique to use in all situations. LCA typically does not address the economic or social aspects of a product.

> Because all techniques have limitations, it is important to understand those that are present in LCA. The limitations include the following:
>
> - The nature of choices and assumptions made in LCA (e.g. system boundary setting, selection of data sources and impact categories) may be subjective.
>
> - Models used for inventory analysis or to assess environmental impacts are limited by their assumptions, and may not be available for all potential impacts or applications.
>
> - Results of LCA studies focused on global and regional issues may not be appropriate for local applications, i.e. local conditions might not be adequately represented by regional or global conditions.

- The accuracy of LCA studies may be limited by accessibility or availability of relevant data, or by data quality, e.g. gaps, types of data, aggregation, average, site-specific.

- The lack of spatial and temporal dimensions in the inventory data used for impact assessment introduces uncertainty in impact results. This uncertainty varies with the spatial and temporal characteristics of each impact category.

Generally, the information developed in an LCA study should be used as part of a much more comprehensive decision process or used to understand the broad or general trade-offs. Comparing results of different LCA studies is only possible if the assumptions and context of each study are the same. These assumptions should also be explicitly stated for reasons of transparency.

This implies that LCA should be used carefully. It is only one of several environmental management techniques (e.g. risk assessment or site related environmental auditing, environmental impact assessment, etc.) and may not be the appropriate tool to use in all situations.

In most situations, it is not possible for a life cycle assessment to prove conclusively that, for example, one product is 'better' than another but this does not mean that it is a waste of time doing it at all. Much can be learned from the process.

The use of computer models in LCA

As you will have seen from the above discussion, carrying out an LCA is a complex and time-consuming task. This can be eased considerably by using one of the many computer models which are currently available. The following computer activity is designed to give you a guided tour of the functions of one of the available programs. It is called Sima Pro and you have been provided with a demonstration version of the software as part of the course. The activity also demonstrates some of the limitations of life cycle assessment.

COMPUTER ACTIVITY 4

In this activity you will explore the functions of Sima Pro. The activity is described in detail in the T308 Computing Guide and the model you need can be found in the modelling section of the Resources DVD.

LCA-based waste management models

There are a number of waste management models which are based on LCA methodology including those developed by the United States Environmental Protection Agency and the CSR/EPIC model from Canada. Two systems of direct relevance to the UK are considered below.

The Environment Agency developed the WISARD (Waste Integrated Systems Analysis for Recovery and Disposal) in the mid 1990s. This was to help inform the development of waste management policy; in particular:

- national policies and the development of Waste Strategy 2000 (DETR, 2000)

- waste management planning at the regional level.

Much of the work done in producing WISARD involved obtaining, reviewing and evaluating the environmental impacts of different waste management processes (from collection through to final disposal). It also included the impacts (both positive and negative) of the reprocessing of materials collected for recycling. WISARD includes a large generic database which allows it to take account of (for example) the impacts due to the production of cement for building an incinerator and the impacts from the manufacture of wheeled bins. This generic database was developed by Ecobilan in France.

WISARD has been used by national regulators such as the Environment Agency, the Scottish Environment Protection Agency and the Government of the National Assembly of Wales and also by a number of local authorities to support development of their waste management strategies.

At the time of writing (autumn 2006) the Environment Agency is in the process of producing a replacement for WISARD. Unlike WISARD, this package will include data on 'novel technologies' such as pyrolysis and mechanical biological pretreatment.

The detergents and cosmetics company Procter and Gamble developed an interest in LCA due to the fact that virtually all its product packaging enters the waste stream at some point. The company established a Global Integrated Solid Waste Management Team which led to the creation of the 'IWM' LCA model. IWM is now in its second edition, 'IWM-2' (McDougall et al., 2001). This is the model that I used to generate Figure 3 in Section 2.

IWM-2 is less complex than WISARD in that it does not consider the impacts due to the manufacture of the waste management plant and equipment. However, the errors arising from this simplification are extremely small. Also, IWM-2 requires less specific data than WISARD and its simpler structure makes it easier to use as a screening tool that provides a quick comparison of alternative strategies.

8.1.3   Multi-criteria analysis and multi-criteria decision analysis

Multi-criteria analysis (MCA) and multi-criteria decision analysis (MCDA) are techniques which can be used to evaluate different options, and they are being increasingly used in the UK and other countries in public decision-making.

Practical guidance in the use of the techniques has been produced by the former DETR and File Article 6 is an edited extract from the manual. For the purposes of this course, the extract has been limited to cover the basic principles and the techniques which you will be able to apply in Section 8.4. If you are interested in finding out more about the techniques you can obtain a complete copy of the guide which (July 2004) can be downloaded from the World Wide Web. A link to this is available on the course website. You will have an opportunity to apply what you have learned about MCDA in Computer Activity 6 in Section 8.4.

READ

Study File Article 6 in Appendix 1 now.

8.2   Introduction to waste management strategies

8.2.1   Integrated solid waste management

In the past, the various options for dealing with municipal solid waste were seen as competitors. A local authority would consider developing either landfill or incineration and possibly include some recycling to demonstrate their 'green credentials'. This situation has now changed and, as a result of the legislative changes, Waste Disposal Authorities are working with their constituent Waste Collection Authorities to adopt a more integrated approach. The term 'integrated solid waste management' (ISWM) simply means treating each part of the waste stream using the most appropriate method while keeping the total negative environmental impacts to a minimum. For example:

■ clean office paper, newsprint and other grades of paper can be recycled in areas where there is a market for this material

■ garden and municipal parks waste can be composted if a local market is available for the compost

■ material that cannot be recycled can be burned to recover the energy in the form of power, heat or combined heat and power

■ inert wastes (such as building waste) can be used in land restoration projects.

Many developed countries are developing ISWM as their national strategy. Examples from the EU are given in Table 39 based on 2003/04 statistics. The important point from this table is that high recycling rates and high energy recovery rates can be achieved at the same time. The minimisation of landfill while maximising the recovery of materials and energy can be regarded as one of the key aims of ISWM.

Table 39   Waste management in EU-15 countries

Country	Municipal solid waste disposal 2003/04 (%)		
	Recycling and composting	Energy recovery	Landfill
Austria	59	11	30
Belgium	52	36	13
Denmark	41	54	5
Finland	28	9	63
France	28	34	38
Germany	57	23	20
Greece	8	0	92
Ireland	31	0	69
Italy	29	9	62
Luxembourg	36	42	23
Netherlands	64	33	3
Portugal	4	22	75
Spain	34	7	59
Sweden	41	45	14
United Kingdom	18	8	74

Source: DEFRA (2006).

8.2.2   Deciding on a strategy

Waste strategies should be firmly based on the principle of BPEO as discussed above. This allows for local circumstances to be taken into account fully. However, in practice, the effect of national policies (such as the BVPI recycling targets for English authorities) place severe restrictions on the scope of local strategies.

The capital investments required to provide facilities for the management of waste materials are considerable, whether for waste to energy plant, materials recovery, waste transfer or the provision of well-engineered landfill sites. The operating and transport costs associated with them are also substantial, while suitable sites, particularly for landfill, can be difficult to find in places where they would be most useful. Good long-term strategic planning is therefore essential if these resources are to be properly managed, and if the environmental impact of waste management activities is to be minimised.

A good waste management strategy will be cost-effective, with an emphasis on the conservation of basic materials and energy. It will be sufficiently flexible to allow for both operational emergencies and uncertainties in forecasts of future waste arisings and predicted costs. At the very least, it should specify:

■ the facilities to be provided for the recycling, recovery, treatment, transfer or disposal of wastes

■ where and when they are to be built

■ what their capacity should be.

Increasingly, the trend is towards waste minimisation and recycling as the highest priorities in a waste management strategy.

The preferred strategy should only be chosen after a number of different options have been fully evaluated and compared, taking into account economic, technical, environmental and political factors, as well as the use and conservation of resources.

Arriving at a suitable strategy, therefore, requires the integrated planning of many different activities, and it should address itself to three main tasks:

(a) reduction in the quantities of waste arising;

(b) recycling or reclamation of as much of the waste as is possible;

(c) ensuring that the remaining material is disposed of according to the best practicable environmental option.

It is essential that basic policy issues should be clearly established including those set at the national or the European level.

Having established the overall objectives, the next stage in developing a strategy is to carry out a study, which should include:

1   analysis of the quantities, composition and distribution of waste arisings;

2   forecasting of future waste arisings (this will depend on changes both in composition and population);

3   assessment of the potential for avoiding the production of waste;

4   assessment of the reuse or recycling potential of the waste;

5   assessment of reclamation potential of the waste, including incineration with energy recovery, and novel recovery options;

6   assessment of available technologies for the above and their economics, reliability and flexibility to cope with future changes;

7   consideration of possible future technologies which might be available within the period of the strategy;

8   assessment of the availability of land for transfer or treatment plants;

9   assessment of the availability of suitable airspace for landfill of various waste types, with particular reference to land which could benefit from being reclaimed;

10   assessment of transportation methods available and routes that could be used;

11   assessment of the preferred location of possible facilities in relation to waste generation areas;

12   assessment of planning and constraints imposed by the national regulator.

All the available options having been assessed, a shortlist of those that are particularly applicable to the local area should be drawn up, and then assembled in different combinations to generate a number of alternative strategies for the area and time period being considered. Each of these should then be evaluated.

Evaluation of even the simplest strategy can be difficult, and the use of computer models to help in this is growing, particularly in calculating the overall costs of each strategy. Such models also offer the possibility of sensitivity analyses for those areas where uncertainty is the greatest. You will be using a computer model which will allow you to make comparisons between different waste management options in Sections 8.3 and 8.4.

The final choice of the preferred strategy for the area should not be based on cost alone. It will need also to take into account the environmental safeguards as well as political factors such as public acceptability and effect on the general amenity of the area. You will be able to take account of such factors in the activity you carry out in Section 8.4.

8.3   What collection, recycling and recovery rates can we achieve?

Chapter 2 of T210 Block 4: *Wastes management*, the Legislation Supplement and Section 1.2.1 of this block discuss the Landfill Directive targets and targets for recycling and recovery which the governments and assemblies of the UK have produced. The targets for England are summarised in Table 40.

Table 40   Municipal waste-related targets for England

Landfill Directive	Reduce the amount of biodegradable municipal solid waste (MSW) sent to landfill to: 75% of the weight produced in 1995 by 2008 50% of the weight produced in 1995 by 2013 35% of the weight produced in 1995 by 2020
Household waste recycling and composting	Recycle or compost at least the following percentages of household waste: 25% by 2005 30% by 2010 33% by 2015 Interim targets for 2003/04 and 2004/05 set for each authority
Municipal waste recovery (recycling, composting and incineration with energy recovery)	Recover at least the following percentages of municipal waste: 40% by 2005 45% by 2010 67% by 2015

Achieving all three of these targets is a complex task and requires an integrated approach.

In this section of the block you will carry out an extensive computer activity using a spreadsheet model in which you will investigate what can be achieved using different combinations of the main waste management methods, namely:

- materials recovery
- composting
- incineration with energy recovery
- landfill.

COMPUTER ACTIVITY 5

The activity is described in detail in the T308 Computing Guide, and the spreadsheet model you need can be found in the modelling section of the Resources DVD.

8.4 Exercise on integrated solid waste management

In this final section of the Waste block you will be developing a waste strategy for Barsetshire County Council for the collection and disposal of the county's municipal solid waste (MSW).

The overriding problem faced by the council, in common with many other districts and counties throughout the UK, is that its landfill sites, which have been taking domestic waste for many years now, will soon be full.

In addition to considering the most appropriate technologies to be used, the *economics* of the strategy will need to be considered, as will the necessary *environmental* safeguards.

Your strategy can include elements from some or all of the waste management options you considered in Computer Activity 5.

You will need to remember that the council has a legal *requirement* to adopt a waste management strategy that will comply with the European Landfill Directive in a cost-effective way without exporting the waste to a neighbouring county. In addition, the council would *like* to meet the *recovery* and *recycling* targets set out in the Government's Waste Strategy.

COMPUTER ACTIVITY 6

The activity is described in detail in the T308 Computing Guide. You will need to use the Integrated Solid Waste Management model to provide the information you will be using in this activity. However, the multi-criteria decision analysis (MCDA) work does not require the use of a computer model.

ANSWERS TO SELF-ASSESSMENT QUESTIONS

SAQ 1

Several factors are contributing to the increase in MSW production, but the principal one is the increase in numbers of households due to increased lifespan, smaller families, more single people living alone. Other factors include:

- increased use of pre-packaged meals
- increasing affluence
- growth in second homes and weekly commuters
- increased provision of wheeled bins.

Also, providing people with a garden waste collection service increases recycling rates due to composting, but also increases the total amount of waste generated because people are less likely to carry out home composting in these circumstances.

SAQ 2

Civic amenity (CA) sites are intended to take waste that cannot be collected as part of the refuse collection vehicle (RCV) collection. So CA waste contains higher proportions of DIY waste (mainly concrete, stone/brick, etc., wood and plastics), bulky items (furniture, electrical and electronic goods) and garden waste (grass mowings, prunings, soil, stones, etc.).

CA waste also contains higher proportions of large packaging materials (cardboard boxes), broken toys and household items (comprising many materials, but mainly metals).

SAQ 3

Waste is an extremely heterogeneous material and, whilst a rigorous process was followed to obtain a representative sample for analysis, wide variations must be expected. For example, all it takes is the presence of (say) one mercury battery in one metal sample, but not in the other sample, to give these order-of-magnitude differences.

These data demonstrate that all waste analyses, and trace chemical analyses in particular, must be treated with great caution. In calculations always present the range of figures to let the reader know the degree of uncertainly and resist the temptation to take average values that would hide this variation.

SAQ 4

There are several reasons for the higher recycling rates of commercial and industrial wastes:

- the material is produced in a much more concentrated form
- the composition and quality of the waste is much more closely defined
- the organisations have a financial interest in recycling if it saves money
- it is good publicity
- the packaging regulations mean that industry is obliged to either fund recycling ventures through the purchase of packaging recovery notes (PRNs) or carry out recycling itself.

SAQ 5

Composting: Health impacts

I would expect dust and bioaerosols to be high on your list, if not in first position. Noise, ammonia and VOCs will probably be next for consideration. Depending on the process details, manual handling could also be an issue.

Composting: Environmental impacts

Again dust and bioaerosols should be high up on your list. Ammonia emissions may also be important depending on local circumstances. Nuisance issues such as noise and odour may also be significant at the local level.

You may also have made a case for including CO_2 emissions. However, it can also be argued that composting is CO_2 neutral because the process merely releases CO_2 that has been extracted from the atmosphere by photosynthesis.

Incineration: Health impacts

Again the main health impacts relate to the exposure to dust, bioaerosols and noise of process workers and, to a much lesser extent, people working or living in the neighbourhood.

You have probably also considered exposure of the wider population to the emissions in the flue gases. This subject receives much publicity but the concentrations experienced and their effects are extremely small. In practice, any health effects cannot be detected in the population. Of these emissions, oxides of nitrogen would have the greatest theoretical impacts.

Incineration: Environmental impacts

The main environmental impacts are the flue gas emissions and, to a lesser extent, the potential for leachate from landfilled incineration residues.

As in the case of composting, the issue is not clear cut when it comes to CO_2. Some of the CO_2 released was formed from photosynthesis and the heat recovery reduces the emissions from the fossil fuels that would otherwise be burned.

SAQ 6

The exposure routes for bioaerosols include:

- inhalation – breathing via nose or mouth
- ingestion – eating or swallowing
- absorption – through skin or via the eyes (directly or via contaminated surfaces/clothing)
- contact – with the surface of the skin or eyes
- injection – by high-pressure equipment/contaminated sharp objects.

SAQ 7

Setting exposure limits for micro-organisms is complicated as:

- there are many natural sources of micro-organisms
- we do not have standardised methods of sampling and enumerating them – indeed there are many different methods available
- satisfactory dose-response data are not available
- there are many species of micro-organisms, all of which may have their own specific health effects at a variety of different concentrations.

SAQ 8

Environmental issues associated with transfer stations include:

- noise from machinery and traffic
- aesthetic issues regarding access roads, e.g. mud and dirt
- odour
- air quality issues, particularly nuisance dusts but also exhaust fumes
- litter
- vermin and birds.

SAQ 9

Potential hazards from MRFs include:

Physical	Chemical	Biological
Manual handling	Hazardous waste residues	Airborne micro-organisms
Ergonomics	Hazardous waste vapours/	Contaminated sharps
Accident, transport, fire	aerosols	Contaminated sharp edges
Noise and vibration	Heavy metals, e.g. lead,	Total and respirable dust
Electromagnetic	mercury	
frequencies	Volatile organic compounds	

SAQ 10

Organic materials consist of 'green wastes' such as leaves, grass cuttings, soft prunings, unwanted topsoil, turf, shredded woody material, etc. and 'kitchen wastes', such as the vegetable peelings and leftovers from meals. However, other items such as paper and card could also be viewed as materials that could be composted.

SAQ 11

Benefits from making and using composts include:

- diversion of methane-forming wastes from landfill
- conservation of peat beds and their associated ecosystems
- possible reductions in the need for energy intensive fertilisers by growers.

SAQ 12

Health and safety issues on a composting site include:

Physical	Chemical	Biological
Manual handling	Hazardous waste residues	Airborne micro-organisms
Transport	Hazardous waste vapours/	Dusts
Accident, fire	aerosols	Contaminated sharps and sharp
Noise and vibration	Heavy metals, e.g. lead,	edges
from plant	mercury	Rats
	Volatile organic compounds	

SAQ 13

There are several categories of sensitive receptor that could include:

- houses and residents nearby
- trade premises
- factories
- schools, hospitals and other public buildings

- public footpaths
- other amenities
- crops and livestock.

SAQ 14

Firstly, we need to calculate the surface area of the windrow.

The width of the sloping side of the windrow (w) can be found using Pythagoras' theorem:

$$w = \sqrt{2^2 + 2^2}$$
$$= 2.83 \text{ m}$$

so the total surface area is

$$2.83 \times 100 \times 2 = 566 \text{ m}^2$$

Emission rate = flux $\times$ area

$$= 3 \times 566$$
$$= 1698 \text{ mg h}^{-1}$$
$$= 1698 \times 24 \times 90$$
$$= 3\ 667\ 680 \text{ mg per cycle}$$
$$= 3.7 \text{ kg per cycle}$$

SAQ 15

For this plant, an emission concentration of 1 mg m^{-3} gives a maximum short-term ground level concentration of 0.028 µg m^{-3}. The emissions limits for SO_2 and HCl are set at 50 and 10 mg m^{-3} respectively, so the maximum ground level concentrations would be

For SO_2: 0.028 $\times$ 50 = 1.4 µg m^{-3}

For HCl: 0.028 $\times$ 10 = 0.28 µg m^{-3}

However, you should note that, in practice the emission levels will be much lower than the limits shown in Table 16 so the ground level concentrations will be correspondingly lower.

SAQ 16

(a) chlorine input to the incinerator

$$= 10 \times 1000 \times \frac{0.8}{100}$$
$$= 80 \text{ kg h}^{-1}$$

Flue gas chlorine content

$$= 80 \times \frac{90}{100}$$
$$= 72 \text{ kg h}^{-1}$$
$$= \frac{72}{3600}$$
$$= 0.02 \text{ kg s}^{-1}$$

But this is in the form of HCl, so the HCl flue gas content is

$$0.02 \times \frac{36.5}{35.5}$$

$$= 0.0206 \text{ kg s}^{-1}$$

$$\text{Flue gas HCl concentration} = \frac{0.0206}{16.67}$$

$$= 0.0012 \text{ kg m}^{-3}$$

or 1234 mg m^{-3}

(b) HCl removal required $= 1234 - 10 = 1224$ mg m^{-3}, or $\dfrac{1224 \times 16.67}{1000 \times 1000}$

$$= 0.0204 \text{ kg s}^{-1}$$

Using the chemical equation $Ca(OH)_2 + 2HCl \longrightarrow CaCl_2 + 2H_2O$,

One kmol of lime reacts with 2 kmol of HCl

or, $40 + 2 \times (1 + 16) = 74$ kg of lime reacts with $2 \times (1 + 35.5) = 73$ kg of HCl.

So the theoretical amount of lime required

$$= 0.0204 \times \frac{74}{73} = 0.0207 \text{ kg s}^{-1}$$

But, in reality, we need twice that amount or 0.041 kg s^{-1}.

(c) Scrubbing residue produced (assuming that the water evaporates in the scrubber)

$CaCl_2$ produced $= 40 + (2 \times 35.5) = 111$ kg for each 74 kg of lime reacted

$$= 111 \times \frac{0.0207}{74} = 0.031 \text{ kg s}^{-1}$$

So the total solids produced equals the sum of the $CaCl_2$ formed plus the unreacted excess lime or

$$0.031 + 0.0207 = 0.052 \text{ kg s}^{-1}$$

(d) Allowing for the dust removal in the bag filters

Required dust removal $= 7000 - 10 = 6990$ mg m^{-3}, or

$$\frac{6990}{1000 \times 1000} \times 16.67 = 0.117 \text{ kg s}^{-1}$$

So, total solids production $= 0.052 + 0.117 = 0.17$ kg s^{-1}

(If you had any problems with this SAQ you should reread T210 Block 0: *Chemistry.*)

SAQ 17

Table 41 shows enrichment factors (concentration in the residue/concentration in the waste).

Table 41   Pollutant concentration in combustion residues

Element	Enrichment factor	
	APC residue	Bottom ash
C	0.2	0.2
Si	1.2	7.2
Al	0.4	1.7
Zn	6	2
As	2.3	1
Pb	15	8
Cd	45	2.6
Hg	520	2.5

Not surprisingly, both residues are depleted in carbon (if we assume that loss on ignition in the residues approximates to carbon in the raw waste). The depletion would automatically lead to enrichment in other species.

Taking the APC residue: this is highly enriched in the volatile metals lead, cadmium and mercury and, to a lesser extent, zinc. This is to be expected: the APC residue consists of scrubbing reagent, entrained particles and material that has evaporated from the bed of burning waste and condensed in the cooler scrubbing unit.

The bottom ash is slightly enriched in the volatile metals (due to the concentration effect when carbon and hydrogen are lost) and also more enriched in the non-volatile element silicon. The lesser enrichment in aluminium is due to the aluminium associated with silicon in the mineral components in the waste.

SAQ 18

There would be a number of problems in trying to market this brine:

1   It would contain trace quantities of heavy metals and other impurities;

2   The specification would tend to vary with time;

3   For both the above reasons the only uses would be in low grade applications;

4   The amounts produced are relatively small;

5   Transport costs could be significant;

6   General market resistance to using a 'waste' as a raw material.

SAQ 19

Over the 30-year period about 19 500 t of methane is produced and 90% of this is released in the first 17 years. (As a point of interest, the total amount produced by the site would be about 20 400 t with 90% of this released in the first 21 years). Any gas combustion plant would have to be designed to handle the peak load in year one, but still be capable of operating efficiently and with acceptable levels of environmental emissions at lower outputs. This would suggest the need for modular units – possibly ones that are container mounted that could be transferred to other sites when appropriate. There will be a time when gas combustion with energy recovery becomes uneconomic and, from this point, gas flaring would be necessary.

The contribution of each class of waste is roughly in proportion to the amount in the waste. Not surprisingly, the readily degradable material (RDO) is responsible for the majority of the gas produced in the first four years or so before the moderately degradable material (MDO) takes over. The RDO is virtually fully decomposed by the end of the 30-year period.

If the RDO was to be removed (through a separate collection of food and garden waste) and replaced with an equivalent mass of MDO, the rate of gas production is decreased (17 000 tonnes released in 30 years) and the total release is also lower (18 100 tonnes). However, the period when gas flaring, as opposed to combustion with energy recovery, and other active control measures are required would be longer in this case.

SAQ 20

We begin by assuming that the wind speed is 5 m s^{-1} and Pasquill stability Category D.

For an NO_x emission rate of 1 mg s^{-1}, the model gives a ground level concentration of 0.0047 µg m^{-3} (the value given in the table for $y = 0$ and $x = 400$ m).

In our examples the emission concentrations are 360, 1500 and 250 mg m^{-3}, and converting these into mass emissions we get:

360 mg m^{-3} is equivalent to $360 \times 30 = 10\ 800$ mg s^{-1}

1500 mg m^{-3} is equivalent to $1500 \times 30 = 45\ 000$ mg s^{-1}

250 mg m^{-3} is equivalent to $250 \times 30 = 7500$ mg s^{-1}

So the ground level concentrations at the boundary are:

For 360 mg m^{-3} $C = 10\ 800 \times 0.0047 = 50.8$ µg m^{-3}

For 1500 mg m^{-3} $C = 45\ 000 \times 0.0047 = 212$ µg m^{-3}

For 360 mg m^{-3} $C = 7500 \times 0.0047 = 35.3$ µg m^{-3}

The WHO limit for atmospheric NO_x is 200 µg m^{-3} so, in the worst case, this limit would be exceeded. You should also note that the maximum ground level concentration occurs 800 m from the engine (and 400 m from the site boundary under the worst conditions) and is around three times the level at 400 m.

SAQ 21

All those with the symbol L represent leachate, which could be contaminated.

L_{off} and L_{in} are obvious candidates for pollutants, but also L_{sep} and L_{rec} could be, depending on the rate of movement through the site.

Obviously, the input D_{liq} will be a possible pollutant.

In some cases R_{off}, which is normally clean water, can become contaminated and will need collection and treatment.

Similarly, to a lesser extent, surface water run-on R_{on}.

SAQ 22

(a) One tonne of waste will contain 0.25 tonnes of water.

   If the in-place density of the waste is 0.7 t m^{-3}, then 1 m^3 of waste will contain

   $0.25 \times 0.7 = 0.175$ tonnes of water

   1 t of water occupies 1 m^3, therefore 0.175 t will occupy 0.175 m^3.
   So 1 m^3 waste will contain 0.175 m^3 of 17.5% water.

(b) Similarly, 40% initial moisture content on an undried mass basis is equivalent to 28%.

   So, typical values of initial moisture content vary from, say, 15% to 30% on a volume/volume basis.

SAQ 23

Table 42 shows one way of representing the relationship. Your diagram may differ in detail but should cover essentially the same information.

Initial moisture content	Field capacity	Saturation
15–30%	30–40%	60–70%

Absorptive capacity		Voidage
10–20%		20–35%

SAQ 24

Using Darcy's law

$Q =$ Leachate flow (m^3 d^{-1})

$k = 10^{-9}$ m d^{-1}

$A = 2500$ m^2

$h = 0.8$ m

$d = 1.5$ m

$$Q = k \times A \times \frac{(h+d)}{d}$$

$$Q = 10^{-9} \times 2500 \times \frac{(0.8+1.5)}{1.5}$$

$$= 3.83 \times 10^{-6} \text{ m}^3 \text{ d}^{-1}$$

or $3.83 \times 10^{-6} \times 365 = 0.0014$ m^3 y^{-1}

SAQ 25

(a) Using Giroud's equation

where

$$c = 0.21$$

$$h = 0.75 \text{ m}$$

$$K_s = 1 \times 10^{-9} \text{ m s}^{-1}$$

Using Hall and Marshall's guidance on membrane defects and assuming that all defects are of the maximum size, we can say that flow through each pinhole

$$= 0.21 \times \left(\frac{5}{1\,000\,000} \right)^{0.1} \times 0.75^{0.9} \times (1 \times 10^{-9})^{0.74}$$

$$= 1.04 \times 10^{-8} \text{ m}^3 \text{ s}^{-1}$$

Similarly the flow through each small hole is 1.4×10^{-8} m^3 s^{-1} and the flow through each large hole is 2.2×10^{-8} m^3 s^{-1}.

So the total site flow

$$= 5 \times 1.04 \times 10^{-8} + 2 \times 1.4 \times 10^{-8} + 0.15 \times 2.2 \times 10^{-8}$$

$$= 8.4 \times 10^{-8} \text{ m}^3 \text{ s}^{-1}$$

$$= 0.007 \text{ m}^3 \text{ d}^{-1}$$

(b) If the effect of the membrane can be neglected we need to use Darcy's equation

$$Q = k \times A \times \left(\frac{h + d}{d} \right)$$

where

$$Q = \text{leachate flow (m}^3 \text{ s}^{-1})$$

$$A = 10\,000 \text{ m}^2$$

$$h = 0.75 \text{ m}$$

$$d = 5 \text{ m}$$

$$Q = 1 \times 10^{-9} \times 10\,000 \times \left(\frac{5 + 0.75}{5} \right)$$

$$= 1.1 \times 10^{-5} \text{ m}^3 \text{ s}^{-1}$$

$$= 1.0 \text{ m}^3 \text{ d}^{-1} \text{ (i.e. increased by a factor of 143)}$$

(c) Taking a very simple approach to the water balance equation we can say that, under steady-state conditions,

Rainfall input = leachate outflow

$$\text{Rainfall input} = \left(\frac{40}{1000}\right) \times 10\,000$$
$$= 400 \text{ m}^3 \text{ y}^{-1}$$

Because we have steady-state conditions this also equals the flow of leachate.

$$\text{The leachate escape} = 3600 \times 24 \times 365 \times 1 \times 10^{-9} \times \left(\frac{5+h}{5}\right) \text{ m}^3 \text{ y}^{-1}$$

So we can write

$$400 = 3600 \times 24 \times 365 \times 1 \times 10^{-9} \times \left(\frac{5+h}{5}\right)$$

$$h = 1.34 \text{ m}$$

SAQ 26

(a) The main reed bed was considered necessary to act as a 'polishing stage' to achieve an additional reduction in COD, suspended solids and iron. The final reed bed would also break down the partially degraded pesticides that leave the ozone treatment stage. The advantage of a reed bed over other systems is the low operating cost.

(b) No water treatment will reduce the chloride content of an effluent (note this is why high-chloride wastes, such as incinerator air pollution control residues, will not be allowed to be landfilled under the Waste Acceptance Criteria).

(c) The ozonation stage is designed to oxidise pesticides (such as mecoprop and isoproturon) which might not be fully removed during biological treatment.

(d) The regulator felt able to increase the discharge standards for chlorides and COD due to the results from the fish toxicity testing and the massive dilution factor when the effluent was discharged. Furthermore, experiments with high ozone dosing rates did not achieve an improvement in COD reduction.

SAQ 27

For the case of a minor aquifer, where there is little probability that the water will be used for drinking water abstraction, it would be appropriate to set an EAL of 250 mg l^{-1}. The model output indicates that we can be 95% confident that the maximum concentration will be below about 210 mg l^{-1}. Therefore, the site would be acceptable.

However, for a major aquifer, where drinking water abstraction is either taking place or is likely to occur in the future, the appropriate EAL would be 40 mg l^{-1} and clearly this value is exceeded. Therefore the site would not be acceptable.

SAQ 28

Key assumptions would include:

- Active leachate control will cease sooner rather than later (say 50 years after site completion).
- The liner and cap will degrade, eventually presenting no barrier to leachate escape.
- The aquifer will be used for drinking water abstraction if its size will permit this.
- Probabilistic models should be used with a 95% confidence level.
- A conservative approach should be taken to selecting EALs.
- All predicted pollution events should be taken into account regardless of when they may occur.

SAQ 29

Firstly, you should note that once a potential pollutant enters a landfill, it has three possible fates:

1   Inert species remain in the site over geological timescales.

2   Reactive species may degrade to some extent (oxidation of organic compounds, adsorption on clays and other materials in the ground).

3   Soluble species will eventually be leached out of the site.

Barriers and control measures may reduce the short-term release of pollutants or reduce the concentration in groundwaters, but in the long term all control measures will fail. Therefore, measures to reduce the environmental and health impacts of leachate are limited but could cover:

- locating sites in areas where there is a major natural barrier (the APC residue site shown in the Resources DVD video clip is situated over a 200 m depth of low-permeability clay)
- restricting the materials permitted to be landfilled (see discussion of the Waste Acceptance Criteria in Section 5.1)
- pretreating wastes to remove potential pollutants. Of course, this will only be effective if the form of the potential pollutants can be changed to convert them to an insoluble form. Chlorides (for example) cannot be made insoluble, but could perhaps be discharged into the sea or an estuary provided other pollutant concentrations are sufficiently low.

SAQ 30

Your summary will differ from mine but you should see that life cycle assessments are concerned with:

- products, processes or activities
- quantifying materials and energy use
- quantifying wastes released
- evaluating environmental burdens
- evaluating and implementing opportunities to affect environmental improvements.

Note also that assessment is of the entire life cycle of the product, process or activity (from 'cradle to grave') including:

- extraction and processing of raw materials
- manufacture
- transportation
- distribution
- use/reuse/maintenance/recycling
- final disposal.

APPENDIX 1: FILE ARTICLES

FILE ARTICLE 1: THE NEXT BIG THING

Maris, Anna (2000) 'The next big thing', *Resource*, April.

Waste could become the biggest health risk in the UK, when it should be earning this country billions of pounds and creating 50,000 jobs. All because we have no national infrastructure for recycling.

As the genetically modified food issue and the transport debate are nearing exhaustion, waste is set to become the next big thing on the environmental agenda. The government's plans to build up to 170 energy-from-waste incinerators will force communities to face up to the threat of 'waste parks' on their doorsteps. These developments are coinciding with a renewed awareness in waste reduction and recycling, as Britain becomes more prosperous. The launch of a National Waste Awareness initiative will go some way towards generating an informed public debate.

Britain is paying over the odds for waste management. The cost for dealing with the 30 million tonnes of household waste annually amounts to over £3 billion. This affects the economy of every household, as 20 per cent of the average council tax bill typically goes towards paying for waste management and disposal. If we utilised waste as a resource, we would not only save ourselves money, but also avoid a potentially catastrophic public health problem.

The UK currently recycles 8 per cent of all household waste, compared to Switzerland's 52 per cent [Figure A2]. Most of our remaining rubbish goes into landfill sites. Apart from the environmental damage involved, it often makes poor economic sense to dispose of or incinerate valuable materials. For example, two-thirds of Britain's used aluminium cans (some 36,000 tonnes) are landfilled every year. It costs us £2 million to dispose of the material, which would be worth £23 million if it was delivered to reprocessors. Similarly the UK has 3.5 million tonnes of waste paper available for recovery. The disposal cost is £175 million, but it could produce paper products worth in excess of £2.2 billion.

[Figure A1]   A sustainable waste programme could push British recycling rates to 80 per cent

The problem is funding the collection of these materials. The UK has no organised infrastructure for recycling, and the government has failed to improve upon the waste policies of the Tory legacy. The European Landfill Directive is now putting pressure on the UK to clean up its act, but our current system can only be described as a Catch 22. A landfill tax was imposed to discourage waste going to landfill. This affects local authorities, who are forced to pay £270 million per year in landfill tax for the only waste disposal option available to many of them. Because of the hefty tax bill, local government has virtually no money left for setting up the collection systems needed to start improving the recycling rate, which would reduce the costs of waste disposal. It is not that the money isn't there. The landfill tax raises almost £500 million, most of which is earmarked to reduce employers' National Insurance contributions. If only half of it was put towards building a recycling infrastructure, it would go a long way towards the government's goal of 30 per cent recycling by 2010. It would also be cheaper than building the proposed 170 incinerators.

British dioxin levels – No 2 in Europe

There are a number of misconceptions about recycling in Britain. The debate on whether it is a viable option goes on, in the face of evidence from many other international and domestic communities which are already diverting 40–80 per cent of their waste. The cost of recycling is often closely examined without looking at the system-wide savings on waste disposal, job creation and other externalities. Much attention is also given to whether it is environmentally sound to transport secondary materials across the UK, when we quite happily buy timber from Scandinavia and Canada for paper production.

An investment in intensive recycling is likely to pay off in a number of areas, including the NHS. The bare facts of dioxin pollution from the incineration of waste are chilling. Last year the World Wide Fund for Nature (WWF) reported that two-month-old British infants are taking in 42 times the safe level of dioxins in breast milk. Neurological and gastrointestinal diseases are only some of the effects of exposure to emissions, and these effects are long term. According to a brand new European Union backed study people exposed to dioxin emissions over 20 years ago, are still suffering as a direct result. Neuralgia attacks, chronic headaches and sleep disorders were commonplace, and there was an above average death rate compared to the wider test group.

Britain has the second highest levels of dioxin emissions in Europe, and according to 1997 figures, 80 per cent is derived from waste disposal, primarily incineration. The main reason for this is that waste is not separated prior to burning. In Denmark, waste for incineration is separated to the same standard as for recycling, so PVC and other noxious fractions can be removed. Here, materials, including plastics, aerosols, pesticides, household batteries and computers are burned. The government reassures us that the emissions are safe, although levels considered safe only ten years ago are now proving to be dangerous.

There is a direct conflict of interest between incineration and recycling. Once built, an incinerator must be fed a regular stream of waste in order to function. It is too costly to turn it off if the local waste runs out. Local authorities who sign up for incineration are more than often contractually obliged to supply a certain amount of waste, and there will soon come a point when arguments break out over who owns Britain's waste.

Choking on our own waste

The bias towards incineration in British waste policy is based on short-term thinking. The ease of planning, budgeting and controlling a waste incinerator make it a convenient option to landfill. Although the benefits of recycling are acknowledged by the government and society at large, incineration is favoured through subsidies and taxes, while recycling is challenged on its economic viability.

The only one way we can avoid choking on our own waste is through recycling, composting and waste reduction. A scheme to recycle 25 per cent of the waste in our bins could cost as little as £85 million. It would instantly create 14,500 green collar jobs within waste collection and sorting in the UK and would generate billions of pounds in terms of indirect job creation, decreases in import and the saving of natural resources.

If weekly kerbside collections were implemented throughout the country and backed this with a public information campaign, a recycling rate of 25 per cent could be achieved comfortably within a year. The methods are currently in use by the Community Recycling Network (CRN), an organisation with over 250 community-based and not-for-profit waste management organisations as members. According to Andy Moore, coordinator of the network, the domestic paper, aluminium and other reprocessing markets would match such a development within three years.

The preferred method of collection by the CRN is weekly pick up of kerbside boxes from the doorsteps of each household. In other words, the householder is given a plastic crate to put all their recyclables in, most often paper, card, glass, cans and aluminium foil. In some areas textiles, shoes, car batteries, plastic bottles, used engine oil, fluorescent tubes and spectacles are also collected. On collection, recyclables are source separated by recycling crews into compartments on their vehicles, to avoid contamination of materials. The method gives a near 100 per cent recycling rate, compared to as low as 60 per cent for mixed bag systems, where mixed recyclables are picked up from households and later sorted at a Materials Recycling Facility (MRF).

The network currently reaches over 800,000 households across the UK, about 4 per cent of the total, with householder separated kerbside collections. As CRN outfits are non-profit distributing, the transparency of the community sector promotes accountability and value for money. However, the community sector is sometimes excluded from the tendering process as a result of the choices made by local authorities. The so-called integrated waste management contracts make the job easy for the local authorities, but may not always achieve best value for council tax payers.

The opportunity

The organic waste stream forms a significant opportunity for high intensive recycling. According to government figures, around 21 per cent of our household waste is compostable organic waste, but independent figures suggest this is significantly higher than previously thought. The research was carried out by Network Recycling over three years and shows that organic waste made up on average 48 per cent of the dustbin during the summer months and 33 per cent during the rest of the year. These figures are backed by several independent local authority waste audits including Daventry, Bury St Edmunds and Wye/Brook in Kent, and complement new innovative composting techniques recently available on the market, where organic matter can be turned into compost over a matter of days or even hours. Combining kerbside recycling with organics composting and adding some elements of reuse and reduction could push the UK recycling rate to over 80 per cent [Figure A1]. The only real barrier for implementation is funding.

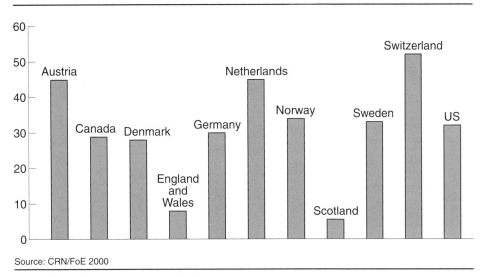

Source: CRN/FoE 2000

[Figure A2]   Recycling figures in Europe

Two major economic initiatives launched by the previous government have failed to channel more money into intensive recycling. Packaging Recovery Notes (PRN), designed to make generators of packaging waste pay for the waste they create to be recycled or have value recovered from it, resulted in £75 million being generated in the first year. However, Britain's recycling rate has not improved as a result of it. It seems that reprocessors such as British Glass and the major paper mills have retained the revenue.

The Landfill Tax Credit Scheme has made 20 per cent of the landfill tax revenue available to approved environmental bodies who work to benefit communities living near landfill sites. Money can also be spent on encouraging the development of more sustainable methods of dealing with waste. Last year £92 million was shared out entirely at the discretion of the commercial operators themselves, without any government input. If this money was earmarked for local authorities to build a recycling collection infrastructure, Britain might stop lining the bottom of the European recycling league table.

In the same way that India now has the opportunity to develop from a preindustrial society into a digital one, without the pollution of an industrial revolution, Britain has the chance to make a huge leap with regards to waste, from landfill to a green economy. Spending up to £100 million of public money on incineration, the waste disposal option next in line for heavy European Union regulation, is a wasted opportunity. Our government now has the power to turn the nightmare of public waste scares into a dynamo of prosperity for Britain.

SUMMING IT UP

Multi-material householder separated kerbside collections, if implemented throughout the country, will have a take-up of 25 per cent of UK dustbin contents. The cost for recycling of 3,593,750 tonnes of the household waste stream amounts to £323,775,000.

As materials are reprocessed, the disposal costs at an average £30 per tonne means that the disposal bill of £107,812,500 can be deducted. Taking into account the effects on the materials markets, as the amount of materials collected will double, an average revenue of £10 per tonne would bring an income of £35,937,500. The net cost of kerbside recycling is then £180,025,000.

If multi material kerbside collection was introduced throughout Britain, 14,500 manual jobs would be created. This would mean that 14,500 people, who currently receive £2,500 in social security benefits, would get back into work, saving the government £36,250,000. The added value of a job in recycling collection would bring an income of £4,000 per job in tax and national insurance payments to the state amounting to £58,000,000. The cost of recycling is then £85,775,000.

FILE ARTICLE 2: BURN ME

Pearce, Fred (1997) 'Burn me', *New Scientist*, 22 November.

Why do we recycle? Frank Ackerman, a professor of environment policy at Tufts University in Medford, Massachusetts, used this question as the title of a book published earlier this year. Recycling, he concluded, 'is one of the most accessible, tangible symbols of the commitment to do the right thing'. In the US, more people recycle than vote. It is 'a religion', writes Ackerman, 'in a society that produces goods far more readily than satisfying beliefs.'

In Britain, one veteran green goes further. Richard Sandbrook, director of the London-based independent research group, the International Institute for Environment and Development (IIED), says that much thinking on recycling is fundamentally misguided because 'environmentalists refuse to countenance any argument which undermines their sacred cow'. He cites new research which turns on its head the conventional wisdom about the recycling of paper and other wastes – wisdom that has become accepted by environmentalists, governments and industry alike throughout the West. The authors of this revisionist work call for a root-and-branch reappraisal of the value of recycling as a means of waste paper management. The message is this: if you value your environment don't put this magazine in the paper bank once you have read it. The green option is to burn it.

Used paper is fast becoming the main raw material worldwide for a business that accounts for 2.5 per cent of all industrial output. In much of Western Europe, more than half of all newsprint is already recycled. The Western world's huge paper disposal industry – which each year handles some 130 kilograms per head in Europe and double that in North America – is being rebuilt on the premise that recycling is best. But even as recycling takes over, economists, business analysts and ecologists are starting to ask hard questions.

Hierarchy heresy

One leading inquisitor is Matthew Leach, an energy policy analyst at the Centre for Environmental Technology at Imperial College, London. With his colleagues, he set out to discover the overall environmental cost of the various waste paper management options. In a study published this month in the *International Journal of Environmental Planning and Management*, they conclude that recycling is better than Britain's favourite option, landfilling, but that it is usually markedly worse than incineration. Moreover, says Leach, 'the higher you value the environment, the better incineration comes out'.

This finding comes as a shock to environmentalists brought up with the teachings of Friends of the Earth and Greenpeace. They advocate a 'waste hierarchy', where the best option is to reduce the use of a resource, next best is reuse and failing that, recycling. After these three 'good' options come two 'bads' – incineration and landfilling. Such thinking has also been adopted by Western governments. It is, for example, enshrined in the European Commission's directive on waste management, which in 1994 set a target for 50 per cent of paper waste to be recovered and recycled by the year 2001.

Leach's study analysed five possible fates for waste paper: recycling to make a similar grade of paper, recycling to make a lower grade, incineration with energy generation, composting, and landfilling with the recovery of methane to generate energy. The researchers then explored the economic and environmental gains and losses from each of these methods. These include the benefits of valuable by-products, such as the sale of electricity generated during incineration, but also the hidden environmental costs, so-called 'externalities' such as carbon dioxide, methane, carbon monoxide, sulphur dioxide, nitrogen oxides and particulates. Cash values for these were then added to, or subtracted from, the costs of each method. The assumption is that the cheapest option, after the environmental externalities have been taken into account, is the best disposal route.

The unique feature of Leach's work is 'that he does not assign his own cash values to the externalities'. Instead, he looks at the range of values assigned in other studies by environmental economists. Then, using one set of studies where the values are low and another where they are high, he assesses how different valuations influence the choice of best disposal route. The final analysis is based on around 50 studies by various organisations ranging from the Swedish Environmental Protection Agency to the Argonne National Laboratory in Illinois and the European Federation for Transport.

Not surprisingly, perhaps, economists disagree violently about what cash values to apply to pollutants. This is a very inexact science. Published environmental costs of emitting a kilogram of carbon dioxide range from $1 to more than $50. In some cases, the environmental cost makes up almost half the cost of paper disposal as calculated using Leach's method.

Leach's work has produced remarkable conclusions about the best disposal route for waste paper. Environmentalists would confidently expect that their studies, which tend to give a high value to environmental externalities, would push the balance in favour of recycling. Business analysts, on the other hand, might minimise the value of externalities, assuming that this would undermine the arguments for recycling. Far from it. In Leach's analysis, studies which value externalities at a low level typically suggest that 80 per cent of the waste should be recycled (the remainder, the poorest quality and most contaminated paper waste such as food wrappings, is useless for recycling and is landfilled). But when externalities are given high values, the study concludes that two-thirds of waste paper should be incinerated, with the remaining poorer

End of the line? Sorting and redistribution are energy-hungry aspects of recycling

[Figure A3]

grades of paper divided between composting and landfilling. The lesson is that recycling paper makes economic sense if you downplay the environmental costs. But if you care about the environment, incineration wins.

One reason for this surprising finding, says Leach, is the value of energy generated by incineration. Another is that recycling uses large amounts of energy and creates pollution, especially when waste paper is transported to recycling mills. For example, the Aylesford Newsprint recycling mill in Kent (see Box) receives 30 000 truck deliveries of waste paper a year from right across England, with a total annual journey of more than 4 million kilometres. According to Leach's calculations, this would account for more than 5800 tonnes of CO_2 emissions per year. Then there are the trips made by individual recyclers from their homes to the neighbourhood recycling bin. One study in rural Norfolk found that cars travelled 270 kilometres for every tonne of waste posted in local bins.

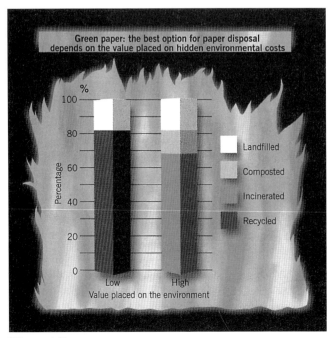

[Figure A4]

Ever wondered what happens to your old papers after you post them through the slot in the recycling bin? In Britain, one in five recycled newspapers and magazines ends up in the Kent village of Aylesford. Here, at Europe's largest recycled newsprint mill, 450 000 tonnes of newspaper is treated annually.

The scale of the operation is impressive. The plant receives 30 000 truck deliveries each year. Once inside the mill the paper is first soaked in water to make a pulp, treated with soap and solvents to remove ink, and then screened, spun and further treated to remove staples, plastics, glue and grit. Cleaned-up pulp is finally reformed into paper on a colossal machine that turns out a constant stream of newsprint 9 metres wide at a speed of 100 kilometres per hour.

Almost one in every 20 tonnes of newsprint used in Europe comes from this one plant. It feeds Rupert Murdoch's presses at Wapping in east London, and many others in Britain, Scandinavia and Germany. On average, the complete cycle from mill to printing press, newsagent, breakfast table, recycling bin and back to the mill takes about 14 days.

Industrial irony

On top of all the motor fuel, the recycling process itself uses energy. The Aylesford plant, for instance, used 4000 tonnes of heavy fuel oil and 5700 million megajoules of gas last year, even with on-site energy and heat recycling. The de-inking process is particularly energy hungry and ultimately produces a toxic sludge containing high concentrations of heavy metals, which must be placed in landfill sites.

Incineration, by contrast, generates energy. Most modern plants have their boilers hooked up to turbines and feed the national grid. The local government's Edmonton incinerator in north London, for example, can produce as much electricity as a 20-megawatt power station for the national grid. Some plants, particularly in Scandinavia, also supply waste heat to neighbouring offices, homes or factories. As with recycling plants, waste is mostly delivered to incinerators by road. But because incinerators also handle much of the rest of domestic refuse, Leach says that they are usually sited closer to the waste source and so generate far less traffic.

True, incinerators produce air pollution. Dioxins, created when some chlorine compounds burn, excite environmental campaigners. But tough new rules on incinerator emissions were introduced in Britain last year, and Leach accepts the view of the Royal Commission on Environmental Pollution that these will effectively eliminate the impact on health of incineration. CO_2 too is often seen as an evil by-product of incineration. If all British municipal waste (including paper) were incinerated, this would increase national emissions of the greenhouse gas by some 3 per cent. But, says Leach, if the wood from which the paper was originally made was replaced by new trees, then those trees would suck up the same amount of carbon dioxide as is emitted when the paper is burnt.

And the evidence is that most paper is produced this way. Contrary to popular myth, only 1 per cent of paper worldwide comes from tropical rainforests. About two-thirds comes from pulp plantations or heavily managed natural forests, mostly in industrialised countries. The big four paper and board exporters are Canada, Finland, Sweden and the US. In all these countries, forest cover is increasing. In Finland's forests, for instance, annual growth is estimated at 85 million cubic metres – some 30 million cubic metres more than the depletion through logging and natural losses. As a result, the incineration cycle is 'CO$_2$ neutral'.

Green gains

What's more, if incineration is then used to generate energy it leads to a net gain for the environment, because it substitutes for the pollution caused by burning fossil fuels, such as coal and gas, in conventional power stations. Leach estimates that every tonne of paper incinerated in the UK saves roughly 300 kilograms of CO$_2$ emissions.

Set against that should be the fossil fuels burnt to produce new paper. Leach agrees that production of virgin paper uses about a quarter more energy than recycling old paper. But he points out that 'most modern Scandinavian pulp mills, from which Britain gets most of its paper, do not burn fossil fuels at all. They burn wood chips and bark.' This fuel is regrown locally, so these mills make no net contribution to global warming. Taking this into account, Leach reckons that 'in terms of fossil energy used to supply a tonne of paper in the UK, virgin paper accounts for roughly half as much energy as recycled paper', leaving aside the extra fuel burned in collecting paper for recycling.

Leach's findings about waste paper management are echoed in a report published last year by the IIED. The work reviews a series of economic and environmental studies of the life cycle of paper, and concludes that 'most of the available studies find that, in some circumstances, incineration can have environmental advantages over recycling'.

The IIED report also warns of the environmental dangers of reducing consumption of virgin paper. It is true that some forestry practices, even in environmentally minded Scandinavia, are far from perfect. But, says the report, a decline in demand for new wood is likely to lead to falling standards of forestry management. After all, why look after a forest you may never be able to profit from. In Finland, timber companies are already allowing plantations to die off, and they blame the policy on increased levels of recycling within the European Union.

So why do governments and environmentalists continue to push recycling? Friends of the Earth has the longest track record of support for recycling. It made its first headlines 25 years ago by dumping nonreturnable bottles on the doorstep of their distributor, Schweppes, and demanding that they be recycled. In those days, Sandbrook was one of the pressure group's three full-time officials. Now, as director of the IIED, he says 'our study concludes that certain green campaigns were plainly misguided'.

And he finds himself at loggerheads with Friends of the Earth's current waste campaigner Anna Thomas. Her latest briefing on waste for local activists rejects the IIED's findings on the grounds that life-cycle analyses 'may be oversimplified or do not use adequate data'. The briefing goes on to back recycling against incineration because the latter causes pollution, 'wastes' a valuable resource, tends to encourage the production of waste and, in a splendidly circular argument, 'represents a barrier to increased recycling'.

Governments seem to have followed the environmentalist line almost blindly. 'The EU's waste hierarchy was never based on any analysis, but was more an article of faith,' says David Pearce. He was a British government environment adviser in the late 1980s and early 1990s, when many of the key decisions were being taken, and is now director of the Centre for Social and Economic Research on the Global Environment at University College London.

Outright opposition from environmentalists, meanwhile, is undermining plans for new incinerators and efforts to hook existing plants up to power generators. Such opposition helped persuade PowerGen, Britain's second largest power generating company, to announce last December that it was pulling out all investment from 'waste-to-energy' schemes. 'This leaves a gaping void of investment into new incinerators,' wrote one commentator in the industry journal *Waste Manager*. In particular, PowerGen has scuppered plans to generate power from a proposed incinerator at Belvedere in southeast London and to convert an existing power station at Kingsnorth in Kent to burning municipal waste.

Deafening silence

As Britain and much of the Western world lurches on towards what Sandbrook describes as further massive investment in ill-conceived recycling schemes, the implications of the new findings need to be thought through carefully. They certainly extend far beyond the paper industry. Work on the life cycles of other wastes, carried out by Pearce's researchers, has concluded that large transport distances to recycling plants, plus the energy used in sorting waste and distributing the recovered materials, 'can quickly undermine the benefits of recycling' [Figure A5]. In many cases, for noncombustible waste, the best disposal method may be landfilling.

Long haul: recycling mills are usually out of town

'Governments seem to have followed the environmental line almost blindly'

[Figure A5]

Leach says that 'reusing glass bottles can use more energy than initial manufacture, since dirty bottles have to be sterilised'. Since glass is made from sand, one of the world's most abundant resources, a better use for old bottles might be to crush them and combine them with other materials to make aggregate.

Yet anyone who questions the sacred cow of recycling risks opprobrium or, perhaps worse, a deafening silence. Sandbrook complains that over the past year, environmental groups have all but ignored his IIED report. This summer, in an attempt to provoke a response, he hit out in the influential environmental journal *Green Futures*. 'One wonders sometimes,' he wrote, 'if the environmental lobby really cares about the efforts being made to get to grips with sustainability.'

The result? 'More silence,' says Sandbrook. 'I really don't know if green campaigners agree with it, disagree with it, or are just bored by it.'

Recycling is bad for the environment but good for your soul. If you have data that contradict this view or ideas about how recycling could be made greener, join the online recycling debate brewing on *New Scientist*'s Web site at http://www.newscientist.com

FILE ARTICLE 3: REMEDIATION OF LEACHATE PROBLEMS AT ARPLEY LANDFILL SITE

Robinson, H., Farrow, S., Last, S. and Jones, D. (2003) 'Remediation of leachate problems at Arpley landfill site, Warrington, Cheshire, UK' in *Proceedings Sardinia 2003, Ninth International Waste Management and Landfill Symposium*, Cagliari, Italy.

Summary

When Waste Recycling Group (WRG) took over Arpley Landfill in Spring 1999, it inherited a site where leachate was up to 8 metres above consented levels. After tankering over 200,000 m^3 of leachate for treatment off-site, investing nearly £3 million in leachate extraction infrastructure and a state-of-the-art on-site leachate treatment plant, discharging into the River Mersey, the site is being transformed. The plant was commissioned in October 2001, and has already treated nearly 150,000 cubic metres of exceptionally strong leachate, to very high standards suitable for discharge into the River Mersey. Dramatic reductions in leachate levels have been achieved in all areas of the landfill site, with the great majority of locations either within, or close to, compliant levels. This paper describes how the project has been implemented, and presents detailed operational data for the leachate treatment system.

1   Arpley Landfill

Arpley Landfill Site is located on the south bank of the River Mersey, west of Warrington, on 130 hectares of land, originally used for disposal of dredgings from the Mersey and from the Manchester Ship Canal. Since it opened in October 1988, the site has received more than 10 million tonnes of domestic, commercial and industrial wastes, emplaced to depths of up to 30 metres, and continues to accept about 800,000 tonnes per year of primarily domestic wastes. Infilled areas are being progressively restored to a mixture of agriculture, woodland and public open space, being blended into the important 186 acre Moore Nature Reserve, which adjoins the site. Wastes have been deposited within three phases, Birchwood, Lapwing and Walton, with increasing degrees of containment. [Figure A6] comprises an aerial view of the site, looking due west.

Tipping cells and phases have variable degrees of basal drainage in place. In the initial Birchwood Phase (18.5 ha, 1988–1994) leachate drainage is by a herringbone of rubble drains. The subsequent Lapwing Phase (40.7 ha, 1994–1997) comprises eight cells with a variety of drainage systems, each leading to a pumping chamber. The current Walton Phase (to be ultimately 45 ha, 1997) has a variety of leachate drainage arrangements, albeit to higher standards than earlier areas.

When WRG acquired Arpley in Spring 1999, from 3C Waste (Cheshire County Council), leachate management at the site was not under control, with no effective infrastructure to enable adequate volumes of leachate to be extracted from the existing chambers.

[Figure A6]   Arpley Landfill, looking due west towards the estuary of the River Mersey showing the leachate treatment plant in centre foreground

2   Initial works at Arpley

2.1 Landfill gas control

Control of landfill gases required high levels of attention and investment at Arpley. Initial works concentrated on improved control of gas emissions – the £8 million Arpley generation scheme is the largest landfill gas power station in the UK, and is capable of generating 16 MW of electricity for export into the National Grid, and was commissioned in October 2000. Gas is extracted from over 200 wells, which had to be drilled and installed, via more than 10 km of pipelines. Also important in control of landfill gas, and of leachate generation, have been works to provide high specification capping to completed areas of filling.

2.2 Leachate collection and extraction

As described, a wide range of drainage systems has historically been installed at Arpley, which did not allow adequate quantities of leachate to be extracted. Extensive investigations by WRG revealed that specific leachate collection chambers were defective, due either to collapse or to silting, and remedial solutions were proposed and implemented. Only by remediating leachate chambers, and by the phased installation of a leachate collection system designed to automatically manage leachate levels across the entire site, could overall yield of leachate be increased and maintained. WRG invested approximately £1 million in the installation of a dedicated pneumatic leachate collection system, together with an interstage storage and transfer compound, all designed with extensive control and management facilities, so as to deliver leachate to a central location, from where it can be directed for disposal. This allowed over 60,000 m^3 of leachate to be extracted and removed during 18 months to the end of 2000. Extended pumping trials during early 2001, demonstrated that extraction of over 300m^3/d was achievable from the combined phases of Lapwing and Birchwood, and that more than 150m^3/d could be pumped from the Walton Phase. Only as the leachate treatment plant has been able to discharge treated leachate off-site, have rates of tankering reduced.

2.3 Leachate characterisation and treatability trials

Enviros was appointed in June 1999, shortly after Arpley began to be operated by WRG, to undertake a review of possible leachate management options at the site. At that time, WRG was negotiating with North West Water plc (now

United Utilities), for disposal of up to 300 m³/d of raw leachate, by pipeline into their nearby Great Sankey STW, across the River Mersey from Arpley. Every indication was being given by NWW and the Environment Agency that this was the favoured option for disposal of leachate, and on this basis WRG invested in a thrust-bored pipeline beneath the Mersey, from the landfill to the inlet of the STW. In fact, this pipeline has not yet been used, as repeated consent failures by the Great Sankey STW, made them unwilling to accept any leachate from Arpley at that time.

The extensive work on leachate extraction at Arpley was meanwhile allowing more representative leachate samples to be extracted and characterised, and better estimates to be made for volumes of leachate, in excess of the licensed maximum 1 metre head, across each phase of the site. Leachates from the different phases were found to vary considerably in quality. Table [A1] summarises their strength in terms of concentrations of chloride, and of ammoniacal-N, one of the main contaminants requiring to be treated.

Table [A1]   Summary of initial data for leachate strength in different phases at Arpley, in terms of chloride and ammoniacal-N (in mg/l)

Landfill phase	Chloride			Ammoniacal-N		
	Minimum	Mean	Maximum	Minimum	Mean	Maximum
Birchwood	38	641	2560	6	274	848
Lapwing	58	2816	12700	8	1048	3380
Walton	592	3522	5500	22	1695	3910

In terms of organic strength, stronger leachates typically contained between 5000 and 10,000 mg/l of COD. BOD_5 values were relatively low – however subsequent work has demonstrated that a significant proportion of the COD is degradable in a well-designed aerobic biological treatment system, with well-acclimatised micro-organisms.

Because the leachates from the different phases vary considerably in their strengths, and because the acceptance criteria for available third party disposal outlets are also different, the leachate collection scheme was designed to separate stronger Walton leachates from the rest of the site. Starting in March 2000, Enviros conducted very detailed leachate treatability studies on leachates from Arpley, using a well-proven pilot-scale Sequencing Batch Reactor (SBR) system in their laboratory, which has been used as a basis for design of several dozen full-scale leachate treatment plants in the UK and overseas (e.g. Robinson and Luo, 1991; Robinson and Strachan, 1999). Leachates from Lapwing/Birchwood, and from Walton Phase, were treated separately. These trials had two main objectives:

(a) to determine whether there were any features of Arpley leachates that are likely to prejudice their successful treatment, either in a well-designed on-site plant, or in combination with domestic sewage at a sewage treatment works; and

(b) to obtain site-specific data that demonstrate the degree of treatment performance that can be anticipated from a full-scale treatment plant, by providing estimates of typical effluent quality, and to allow the treatment process to be optimised.

Results from the trials are summarised below in Table [A2]. Some Arpley leachates are relatively strong and undiluted. At 2000 to 2500 mg/l, concentrations of ammoniacal nitrogen are at least twice those typically found at other sites. Values of COD, or organic matter resistant to biological degradation in treated effluent, were also relatively high, in the order of 2000 mg/l.

Nevertheless, the treatability trials showed that all of the ammoniacal nitrogen and BOD_5 could be successfully treated, without any signs of inhibition, although it was noted that chromium concentrations in Walton leachate were marginal, and might subsequently require specific control and pre-treatment. Low levels of other heavy metals were found unlikely to provide any constraints. At this stage, the presence of low concentrations of PCBs in some Arpley leachates had not become an issue in negotiations regarding off-site disposals, and had played no part in the selection of the leachates for the trials. It was recognised that PCB concentrations in Birchwood and Lapwing leachates were very low indeed, not having been detected in the majority of all samples, since regular PCB monitoring commenced at nearly 40 locations across the site on June 2000, every 3 or 4 weeks.

Table [A2]   Results from pilot-scale treatability trials on Arpley leachates, 2000–2001

Determinand	Walton phase		Lapwing phase	
	1	2	1	2
COD	8260	3160	8510	2880
BOD_{20}	4990	38	1450	57
BOD_5	3020	22	495	14
TOC	3350	717	2500	835
Fatty acids (as C)	639	<10	32	<10
pH-value	8.0	8.8	8.1	8.8
Ammoniacal-N	2110	1.1	2050	0.8
Nitrate-N	1.2	3130	0.5	3310
Nitrite-N	<0.1	0.1	<0.1	0.1
Alkalinity ($CaCO_3$)	11500	1600	11500	2700
Chloride	5200	5060	3150	3180
Chromium	2.17	1.96	1.0	0.82
Nickel	2.81	2.55	1.7	1.01
Copper	0.04	0.05	<0.05	0.09
Zinc	0.33	0.46	0.19	0.32
Lead	0.11	<0.04	0.19	<0.04

Notes: Results in mg/l, except pH-value. 1 = leachate; 2 = effluent

Even in leachates where trace levels of PCBs were detected, typical levels were below one microgram per litre, and Enviros was able to demonstrate that the total amount of PCBs in leachate held within all Arpley phases was about 112 grams. To place this into a realistic perspective, this was calculated to be equivalent to the mean quantity of PCBs flowing past the Arpley landfill every 11 hours, in the River Mersey, during the previous four years (Environment Agency, 2002). In addition, wastewater treatment fate models such as the USEPA 'Water 8' model, and results from the treatability trials, demonstrated that a very high percentage of PCB congeners are removed during biological treatment. The only other significant trace organic compound detected in Arpley leachates was the herbicide mecoprop, ubiquitous in landfill leachates (Environment Agency, 2001), but readily and effectively degraded by biological treatment processes.

On this basis, given that the Great Sankey STW was no longer available as a disposal option for either raw leachates or even biologically pre-treated leachates, and that the continued use of tankers to transport leachates long distances to other STWs was not a sustainable option, a decision was made by WRG to construct a large on-site, leachate treatment plant, capable of reliably meeting the tight discharge consents that would be applied for a discharge of effluent into the River Mersey.

3   The Arpley leachate treatment plant

Following protracted negotiations with local Environment Agency officers, a Water Resources Act 1991 consent was finally forthcoming for a discharge into the River Mersey, which contained the following main numerical limits:

max daily volume: 600 m^3

ammoniacal-N: 15 mg/l

pH-value: 6–9

suspended solids: 45 mg/l

polychlorinated biphenyls: not detectable

BOD$_5$: 30 mg/l

Restrictions were also imposed on a range of heavy metals, and on various trace organic substances, but were not felt likely to threaten successful discharge of treated effluent. In addition, separate pilot-scale treatability trials had been undertaken by Enviros on leachate specifically selected as having the highest single PCB concentration anywhere on the Arpley site (up to a maximum single value of 12 µg/l). These had demonstrated complete removal to <0.1 µg/l, (below the limit of detection), and on this basis, construction of the full-scale plant began in mid 2001. The plant is shown towards the end of construction during October 2001 in [Figure A7], and in an essentially completed state in [Figure A8].

[Figure A7]   Plant under construction, showing the twin raw leachate balancing tanks, three large SBR tanks, and effluent balance tank and reed beds top right

[Figure A8]   The completed plant, spring 2002, showing the
3 roofed SBR tanks, the twin bunded leachate balance tanks,
and the effluent balance tank and reed beds

Results from the treatability trials allowed optimised process and detailed
designs to be prepared for the full-scale plant by Enviros, working in
partnership with civil engineers May Gurney, mechanical and electrical
engineers Hytech Water, and staff of WRG. The plant is designed to treat up
to 300 m^3/d of leachate containing 8000 mg/l of COD and 2,500 mg/l of
ammoniacal-N, or correspondingly higher volumes of weaker leachates, at rates
of up to 450 m^3/d. Treatment comprises aerobic biological treatment in three
very large roofed SBR reactors, with dosing of nutrients and automated control
of pH-value in each tank, following the process designs now incorporated into
more than sixty full-scale plants worldwide. Effluent from each SBR discharges
into an effluent balance tank, before passing through a Dissolved Air Flotation
(DAF) system, which has proved extremely effective in removing residual
solids, and also many colloidal materials.

[Figure A9]   Effluent balance tank, and recently planted reed beds
(October 2001)

Although operation of all functions of the plant is completely automated, and
used a programmable logic controller for maximum reliability, the plant
operator interfaces with this by means of a bank of 4 PCs, one to interrogate
and programme operation of each SBR reactor, and one to coordinate the
operation of the overall treatment system. By December 2001 the treatment
plant biological commissioning phase had been completed, each SBR tank
having been originally 'seeded' using a combination of biomass from existing
full-scale leachate treatment plants, and a local sewage treatment works.

The plant control system contains a great many safeguards and failsafe systems,
to ensure that in the event of any malfunction or equipment failures, no adverse
effects ensue. In such events, alarm functions take over, the operator is called,

and the malfunctions can be identified and a decision made remotely, using a simple telephone link. Operational data are stored by the system, and can be interrogated readily, for example to produce performance data.

Leachates being treated by the plant have been very similar to those anticipated, based on the extensive and detailed programme of sampling undertaken. In practice as large volumes of leachate have continued to be extracted from each phase, to be tankered away or treated, a blended leachate quality has stabilised, and concentrations of PCBs and chromium have reduced significantly. This has avoided any need for the construction of a chromium pre-treatment system for leachates, although land had been set aside in case this was required.

Figures [A10] and [A11] compare COD values in incoming leachates with those in effluent discharged to the Mersey. Mean values have been about 10,000 mg/l from the Walton Phase, and about half of this in leachate from the combined Lapwing/Birchwood Phase. COD values in treated effluent have fallen from initial values of up to 2000 mg/l in early 2002, to very stable levels of about 1000 mg/l since summer 2002. This is probably the result of a number of factors, including continued acclimatisation of the biological sludges which effect treatment, improved experience in selection of flocculants for the DAF plant, establishment of the reed beds, and not least the increasing experience of the plant operator. BOD_5 values in effluent are routinely less than 10 mg/l, and monthly tests for effluent toxicity using the Microtox[R] procedure have never found any detectable toxicity to the sensitive bacteria used in the test.

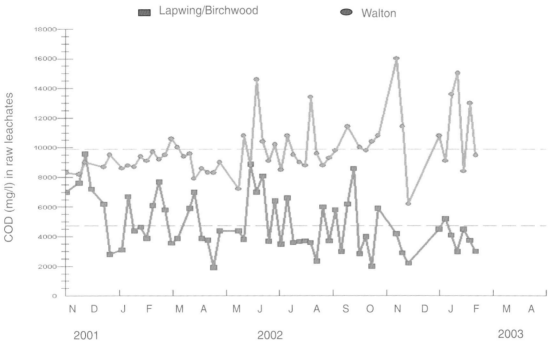

Figure [A10]   COD values in raw leachates extracted from the Lapwing/Birchwood, and Walton Phases at Arpley, November 2001 to March 2003

Figures [A12] and [A13] provide equivalent data for concentrations of ammoniacal-N in raw leachates and in treated effluent. Again, much higher values of up to 4000 mg/l, typically about 3000 mg/l, are measured in Walton leachate, whereas leachates from Lapwing/Birchwood have now stabilised at about 1200 mg/l. The blend of leachates being treated typically contains from 1400 to 1800 mg/l of ammoniacal-N, but effluent values rarely exceed 5 mg/l (Figure [A13]). Removal is primarily by full nitrification – nitrite-N is not detected at significant levels in final effluent.

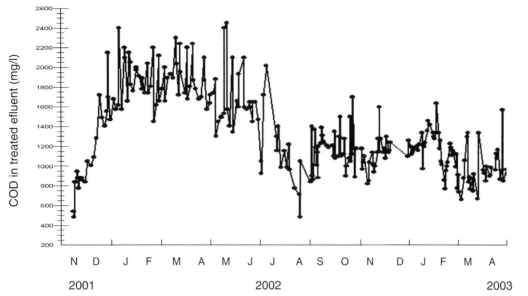

Figure [A11]   COD values in effluent discharged to the River Mersey, November 2001 to May 2003

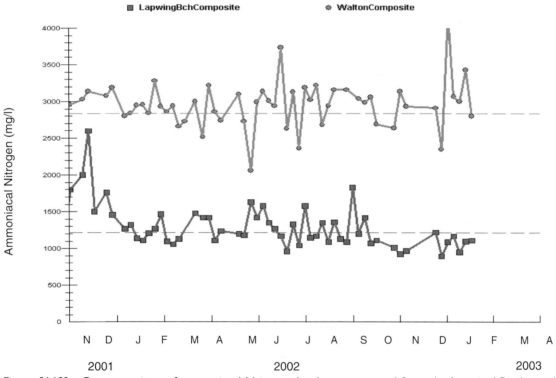

Figure [A12]   Concentrations of ammoniacal-N in raw leachates extracted from the Lapwing/ Birchwood, and Walton Phases at Arpley, November 2001 to March 2003

Table [A3] presents a typical detailed picture of treatment of leachates by the full-scale plant, providing detailed analyses of the raw leachates, the final effluent, and the intermediate treatment stages. Increases in sodium values during treatment result from the use of NaOH, to provide alkalinity needed to maintain pH-values in the optimum range, programmed in the automatic control system. Concentrations of this, and of other salts such as chloride which are essentially untreated by the plant, are not of concern in effluent, because of the tidal nature of the Mersey as it passes the site.

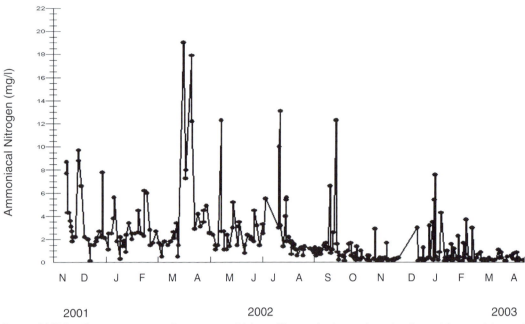

Figure [A13]   Concentrations of ammoniacal-N in effluent discharged to the River Mersey, November 2001 to May 2003

Other contaminants, such as mecoprop, are completely degraded during biological treatment, although present at up to 50 µg/l in raw leachates. PCB congeners have never been detected in effluent (detection limit of <0.1 µg/l), although samples are frequently tested.

Effluent from the DAF plant is polished by passage through a series of terraced reed beds ([Figure A9]) – a wholly natural process in which the reed plant rhizomes provide additional treatment to high standards, as part of Best Available Technology – before final effluent is discharged into the Mersey at typical rates of 10–20 cubic metres per hour. Delivery of leachate into the SBR tanks involves two large and bunded feed storage tanks, allowing effective blending of leachates from the various parts of the site, for optimum treatment.

4   Impacts on the landfill site

Extensive tankering was undertaken from mid-1999, soon after WRG took control of the site, until the treatment plant began to be commissioned in Autumn 2001. This showed the benefit of installing a coherently designed and managed, pumped leachate collection system, which resulted in greatly increased capacity to pump leachate during 2001 and subsequently.

Figure [A14] summarises cumulative removal of leachates from Arpley during the period from early 2001 to May 2003 (63,000 m^3 of leachate having been tankered off-site prior to this period). On-site treatment of leachate has now effectively taken over from tankering, and during this 27 month period a total of more than one quarter of a million cubic metres of leachate has been removed steadily and consistently, at an average rate close to 10,000 m^3 per month.

The impact on leachate levels within the landfill site has been considerable. Figure [A15] summarises changes in mean leachate head in each phase of Arpley – each level representing an average measured depth of leachate, from a large number of monitoring points within that phase.

Table [A3]   Results for treatment of leachate in the full-scale plant at Arpley Landfill, March 2003
(leachate treatment rate at 360 m^3/d)

Determinand	Walton leachate	Lapwing leachate	SBR effluent	DAF effluent	Final effluent
COD	5990	4730	1470	1060	1010
BOD_{20}	1720	1130	67	6	16
BOD_5	688	537	20	<1	<1
TOC	1240	1260	356	281	286
Fatty acids (as C)	26	31	<5	<5	<5
pH-value	8.3	8.2	8.2	7.6	8.1
Ammoniacal-N	1460	1240	3.7	3.2	1.5
Nitrate-N	1.9	1.3	1490	1238	1238
Nitrite-N	<0.1	<0.1	0.6	0.1	<0.1
Alkalinity (as $CaCO_3$)	9110	7490	1430	1010	880
Conductivity (μS/cm)	19700	16700	16300	16800	17500
Chloride	2710	2430	2300	2650	2670
Sulphate	–	113	240	205	197
Phosphate	13.9	9.0	5.8	0.2	0.5
Sodium	2560	2140	3490	3770	3840
Magnesium	93	107	87	89	80
Potassium	1100	886	902	955	991
Calcium	134	173	119	108	100
Chromium	0.67	0.46	0.37	0.27	0.29
Manganese	0.68	0.86	0.18	0.056	0.26
Iron	13.0	12.1	5.51	0.72	0.67
Nickel	0.74	0.61	0.45	0.44	0.46
Copper	0.08	0.029	0.025	0.026	0.038
Zinc	0.29	0.15	0.19	0.19	0.20
Cadmium	0.0020	0.0020	0.0020	0.0010	0.0010
Lead	0.11	0.050	0.028	<0.005	0.009
Arsenic	<0.010	<0.005	0.020	<0.010	<0.010
Mercury	<0.0001	<0.0001	<0.0001	<0.0001	<0.0001

Notes: Results in mg/l, except pH-value, and conductivity (μS/cm); samples taken on 11.03.03

The Arpley Landfill Licence specifies that leachate levels in each phase of
the site shall not exceed 1 metre above the base of individual cells, and it is
clear from Figures A14 and A15 that the massive efforts made by WRG since
they acquired the site have resulted in considerable progress towards this
objective. In each phase of the site levels of leachate are now either compliant,
or close to being compliant – a significant achievement. Continued pumping
and treatment of leachate during coming months is likely to achieve full
compliance during 2003.

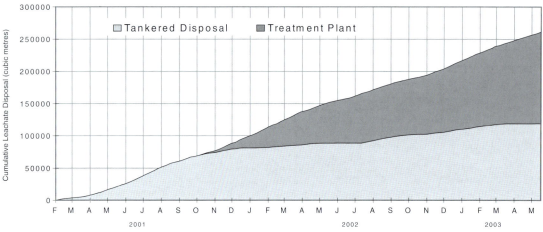

Figure [A14]   Cumulative removal of leachates from Arpley, over 27 months from 2001–2003, for off-site treatment via tanker, or on-site treatment and disposal into the River Mersey (note: 63,000 m^3 of leachate also tankered off-site prior to 2001)

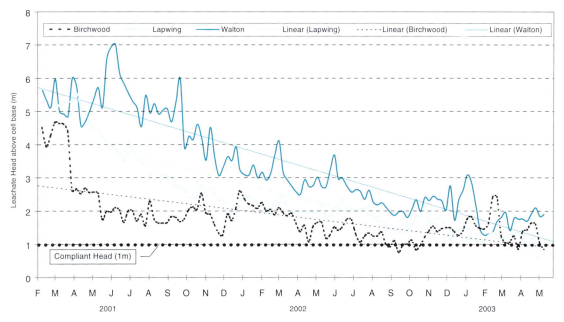

Figure [A15]   Changes in mean leachate head within each phase of Arpley Landfill Site, February 2002 to May 2003 (metres above cell bases)

5   Conclusions

The remediation of serious leachate problems at Arpley, over a four year period since the site was acquired by Waste Recycling Group in Spring 1999, represents one of the largest such projects ever undertaken at a UK landfill site. Leachates across the various phases of the site have been characterised in detail, and with mean COD values in the range 5000 mg/l to 10,000 mg/l, and concentrations of ammoniacal-N from 1000 mg/l to 4000 mg/l, they represent some of the strongest leachates to be reported at a UK landfill.

Extensive engineering works have been undertaken to enable leachate to be extracted at adequate rates from the three phases of the site. The completed pumping scheme has allowed leachate to be extracted reliably and consistently over a 3-year period, at rates of up to 16,000 cubic metres per month, and at a long-term average rate of about 10,000 cubic metres per month.

One of the largest on-site leachate treatment plants in Europe has been commissioned at the Arpley site, following extensive treatability studies, and since October 2002 has treated a total of nearly 150,000 cubic metres of leachate, consistently and reliably, at rates of up to 13,500 m^3/month, enabling safe discharge of treated effluent into the River Mersey.

From a situation where leachate levels were up to 8 m out of compliance when the site was acquired, levels are now compliant, or close to compliance, in all phases of the landfill. Levels are still falling, as leachate continues to be extracted and treated at average rates of about 10,000 m^3/month, and it is anticipated that levels will be brought into compliance across all areas in the near future. The scheme is Best Available Technology, and a state-of-the-art solution.

6   References

Environment Agency (2001) Pollution Inventory discharges to sewer or surface waters from landfill leachates. Report prepared for the EA by Enviros Aspinwall and Knox Associates, ref REGCON 70, May 2001.

Environment Agency (2002) What's in your backyard?, Discharges to Sea-OSPAR. Data for the Mersey Estuary annual discharges of PCBs, held on the Agency website at www.environment-agency.gov.uk.

Robinson H D and Luo M M H (1991) Characterisation and treatment of leachates from Hong Kong Landfill Sites. Paper presented to IWEM '90, 'Design and Construction of Works for Water and Environmental Management', held at the Scottish Exhibition and Conference Centre, Glasgow, 4–6 September 1990, paper 15, 20pp, and published in the *Journal of the Institution of Water and Environmental Management*, 5, (3), June 1991, 326–335.

Robinson H D and Strachan L J (1999) Simple and appropriate landfill leachate treatment strategies in South Africa. Paper presented to 'Sardinia 99', the Seventh International Waste Management and Landfill Symposium, held in S. Marguerita de Pula, Cagliari, Sardinia, Italy, 4–8 October, 1999. In. Proceedings, Volume 2, 269–276.

7   Acknowledgements

Thanks are due to many people who have contributed to the success of the Arpley project. Steve Jennings, Managing Director of WRG's Western Division, and Tim Walsh, Group Technical Director of WRG, have led the project on behalf of Waste Recycling Group, supported by Regional Landfill Manager Mark Kirk and many other colleagues. For Enviros, Keith Corden and Nikki Alexander were particularly involved in detailed design and commissioning of the leachate treatment plant. Alan Loughlin and Dave Horkan led the project team for May Gurney, Mike Beaumont for Hytech Water, and Roger Dixon of Viridian Systems was closely involved in design and installation of the leachate collection and pumping systems. The leachate treatment plant has been operated by Lorna Davies and John Paul Roberts of WRG.

FILE ARTICLE 4: ADVANCED LEACHATE TREATMENT AT BUCKDEN LANDFILL

Robinson, H., Walsh, T. and Carville, M. (2002) 'Advanced leachate treatment at Buckden Landfill, Huntingdon, UK', Institute of Wastes Management annual conference.

Abstract

During April 1994, Aspinwall & Company was appointed to design and commission a treatment plant to treat leachate from the Buckden South Landfill Site, located about 1 km SW of Huntingdon in Cambridgeshire. This formed part of major restoration works being undertaken by AntiWaste Ltd (now part of the Waste Recycling Group (WRG)), following purchase of the site from Hunts Refuse Disposal Ltd. The landfill is located in the flood plain of the river Great Ouse, that flows within 400 m of the site.

Following leachate characterisation and flow studies at the site during early 1994, after an extensive fin drain collection and interception system had been installed, Aspinwall undertook very detailed leachate treatment trials, as part of the detailed design of the on-site treatment plant which needed to be capable of treating 200 m^3/d of leachate to exceptional standards.

The trials were carried out to investigate and design the aerobic biological process that would provide primary treatment, before secondary effluent polishing stages that would involve reed beds and ozonation. The latter specialised stage would be needed to provide treatment of a range of pesticides, including mecoprop and isoproturon, which were found in leachate, and might not be removed fully by biological treatment stages.

The leachate also contained contaminants typical of methanogenic conditions, including ammoniacal-N at values in excess of 400 mg/l, with a required effluent discharge consent standard of 10 mg/l. The treatment plant design had to ensure that these effluent quality standards, including fish toxicity criteria, could be met at all times. The plant was commissioned by Aspinwall during late 1994 and has been operated reliably and consistently by WRG for nearly eight years, primarily based on weekday visits of about 30 minutes by a trained operator. A great deal of data has been accumulated by WRG from detailed monitoring of the plant during this period, including extensive and original data on the effects of ozonation as a leachate polishing process, and this paper provides an opportunity for these to be published for the first time.

The case study described in detail in the paper, represents without doubt the most advanced leachate treatment system in the UK, and one of the most advanced in Europe, and has operated successfully since December 1994. As well as strict restrictions on COD, ammoniacal-N, and usual effluent characteristics, the discharge consent for the plant demands the first toxicity-based standard to be applied to a leachate plant in the UK. This requires that rainbow trout (*Oncorhynchus mykiss*) shall be capable of living satisfactorily for 72 hours in the final effluent that is discharged into the River Ouse.

Introduction

Increases in the extent of regulation of discharges of effluent into the environment (Urban Wastewater Treatment Directive, European Pollutant Emissions Register, nutrient limits, Habitat Directive, etc.) are rapidly imposing more restrictive limits on landfill operators who seek to construct on-site leachate treatment facilities that will make discharges of effluent into surface watercourses. Design of such treatment systems [must] continually advance to meet such discharge limits reliably and cost-effectively. Although knowledge about leachate composition and variations in strength have increased greatly during recent years (e.g. Robinson, 1989, 1996; Knox and Robinson, 2000; Robinson, 2002), many leachate treatment plants continue to be constructed that prove incapable of operating adequately, or of meeting required discharge standards reliably, consistently, and cost-effectively. The case study presented represents a state-of-the-art demonstration of what can be achieved by an appropriate, BPEO solution.

The Buckden project

During April 1994, Aspinwall and Company (now Enviros Aspinwall) was appointed to assist Hunts Refuse Disposals Limited and R T James and Partners, in the design of a treatment plant to treat leachate from the Buckden South Landfill Site, located at NGR TL 207 685, about 1 km south west of Huntingdon in Cambridgeshire. Major restoration works at the site had been undertaken since mid-1994, when the landfill closed. A substantial part of this work required the design and construction of leachate management works, including the provision of a large leachate treatment works adjacent to the landfill site, in the flood plain of the River Great Ouse that flows within 400 m of the site. This plant has proved capable, over nearly 8 years, of treating up to 200 m^3/d of leachate to very high water quality standards. The discharge consent includes a requirement that rainbow trout (*Oncorhynchus mykiss*) shall be capable of living in the final effluent that is discharged into the River Ouse. This almost certainly remains the highest consent standard yet applied to a discharge of treated landfill leachate in the UK.

During the summer of 1994, Aspinwall carried out characterisation studies on leachate from a fin drain collection system at the site. Leachate quality remained relatively consistent during a period from April to June 1994 (see Table [A4]), with the notable exception of mecoprop, where significant and substantial differences between results from different laboratories had to be addressed. Table [A4] also includes indicative consent values originally proposed by the then NRA (Anglian Region) for discharge of treated effluent into the River Great Ouse.

The second part of Aspinwall's work involved detailed leachate treatability trials, that were carried out in order to design and investigate the aerobic biological treatment process that would provide primary treatment of leachate, before secondary polishing stages to involve ozonation and reed beds – mainly for removal of mecoprop and isoproturon herbicides which were identified in the leachate, and might not be removed fully or at all by the first 2 treatment stages. Contaminants that would be of concern, if untreated prior to discharge, were as follows.

COD and **BOD$_5$** values in leachate were well in excess of proposed consent values, but in an uninhibited and acclimatised aerobic biological treatment system would be substantially reduced. Certainly, anticipated BOD$_5$ values in effluent would be well below 20 mg/l, and many existing UK plants have demonstrated such performance over many years. The proposed ultimate COD consent limit of 100 mg/l would, however, prove more of a challenge, without relatively sophisticated additional polishing stages in place. This is in spite of the fact that the compounds which would comprise this COD would comprise harmless products such as fulvic and humic acids, unlikely to pose problems to any aquatic organisms. Residual COD would also be further reduced to some extent by the secondary reed bed treatment stages, and also by the ozonation plant.

Table [A4] Summary of analytical results for leachate pumped from the fin drain at Buckden South Landfill Site, and comparison with proposed consent values received from the NRA (Anglian Region) on 3 June 1994 (d)

Determinand	Minimum value	Maximum value	Mean value	Consent 1 value (b)	Consent 2 value (b)	Consent 3 value (b)
Suspended solids	–	–	–	60	40	40
pH-value	7.1	7.4	7.3	6 to 9	6 to 9	6 to 9
COD	**859**	**1500**	**1093**	**200**	**100**	**500**
BOD$_5$	**50**	**546**	**284**	**40**	**20**	**20**
ammoniacal-N	**260**	**307**	**285**	**20**	**10**	**10**
chloride	**1230**	**1560**	**1455**	**2000**	**1000**	**2000**
BOD$_{20}$	396	1370	915	–	–	–
TOC	180	406	265	–	–	–
fatty acids (as C)	<10	132	66	–	–	–
alkalinity (as CaCO$_3$)	1800	2660	2320	–	–	–
conductivity (μS/cm)	6290	9000	7630	–	–	–
nitrate-N	<0.3	2.0	0.5	–	–	–
nitrite-N	<0.02	<0.1	<0.1	–	–	–
sulphate (as SO$_4$)	20	51	28	3000	1500	1000
phosphate (as P)	0.7	1.3	0.9	–	–	–
sodium	886	1120	1031	–	–	–
magnesium	69	82	76	–	–	–
potassium	130	235	200	–	–	–
calcium	146	256	205	–	–	–
chromium	<0.02	<0.02	<0.0	3.0	1.5	1.5
manganese	<0.04	1.45	0.9	–	–	3.0
iron	**2.9**	**24.5**	**19.6**	**5.0**	**2.5**	**2.5**
nickel	0.07	0.13	0.1	2.0	1.0	1.0
copper	<0.02	0.17	0.0	0.5	0.25	0.25
zinc	0.06	0.60	0.2	–	–	1.0
cadmium	<0.01	<0.01	<0.0[1]	0.02	0.015	–
lead	<0.04	0.57	<0.1	1.0	1.0	1.0
arsenic	0.023	0.058	0.0	0.3	0.15	0.15
mercury (μg/l)	<0.1	<0.1	<0.1	1.0	0.5	0.5
mecoprop (μg/l)	**0.39**	**270 (c)**	**(c)**	**100**	**50**	**50**
isoproturon (μg/l)	**23**	**310**	**205**	**100**	**50**	**50**
Microtox$^{®}$ (per cent)	–	–	–	–	–	**>50%**

Notes:

(a) Results in mg/l except pH-value and conductivity or where shown; – = no result or limit.

(b) Consent values relate to quoted possible initial standards for a discharge of 200 m^3/d of effluent. Consent 1 relates to initial 6 month period of treatment operation. Consent 2 relates to subsequent long term discharges. Consent 3 is the finally-agreed limits, still existing at present.

(c) See text for detailed discussion.

(d) Results are provided in **bold** for determinands present at levels which would cause problems if not treated prior to discharge.

Ammoniacal-N was the main leachate constituent with potential to adversely affect the quality of a receiving watercourse, present at concentrations of up to 300 mg/l, with a required effluent value of 10 mg/l.

Chloride concentrations in leachate were typically mid-way between the values of 2000 mg/l and 1000 mg/l which were proposed as consents for the initial 6 months period, and for the regulation or longer-term discharges respectively. Since none of the treatment stages proposed would reduce concentrations of chloride significantly in any way, and since processes which might (such as reverse osmosis) have serious technical drawbacks, it was necessary to hold further discussions with regulatory authorities. As chloride is a relatively harmless contaminant at levels found, and since considerable dilution is available within the River Great Ouse, this did not cause problems.

Iron was also present at levels which exceeded consented limits, but could readily be reduced to below these by several means, including the aerobic biological system and reed bed which were being proposed.

Mecoprop and **isoproturon** are herbicides which are of concern to the NRA (now the Environment Agency). Before undertaking treatment trials, biological treatment processes could not be relied upon to remove a significant proposition of these pesticides from the leachate – although it was anticipated that mecoprop would be relatively biodegradable. Concentrations of isoproturon consistently exceeded allowable limits of 50 µg/l, by 5 or 6 times, as did mecoprop once analytical discrepancies had been removed by adding a hydrolysis step to include esters of mecoprop in the reported result. A decision was therefore taken that secondary ozonation would be included within the full-scale leachate treatment process at Buckden, in order to ensure that levels of pesticides were adequately reduced.

A further effluent quality standard also specified was that effluent shall not contain any matter which will cause the receiving waters to be poisonous or injurious to fish, the spawn of the fish, or the food of fish. Specifically, 'when tested in accordance with the current OECD Guidelines for Testing of Chemicals 203 – Fish Acute Toxicity Test – the discharge shall have a median lethal concentration over 96 hours (96 hour LC_{50}) of greater than 50 percent'.

It was clear, therefore, that the main contaminant which must consistently be reduced by treatment was ammoniacal-N, where reduction from up to 400 mg/l to below 10 mg/l was essential. Reductions in BOD_5 values from up to 500 mg/l down to below 20 mg/l were also needed but should be readily achieved, and COD values of up to 1500 mg/l must ultimately be reduced to below 100 mg/l unless revised consent conditions could be negotiated. While COD might be reduced further by the ozonation stage, reductions to values approaching the consent level in the biological and reed bed treatment stages would result in significant savings in running costs. The combination of aerobic biological treatment and reed bed polishing should readily ensure that effluent concentrations of other contaminants such as suspended solids, and iron, remained well below proposed discharge limits.

A pilot-scale leachate treatability unit was constructed and used in a similar manner to those which had been successfully in many previous studies (e.g.: Robinson and Luo, 1991). It was operated on a sequencing batch reactor (SBR) basis for nearly 10 weeks, treating about 16 litres of leachate each day, at a controlled temperature of 10 °C (see [Figure A16]). Results for composition of influent leachate, and of settled and filtered effluents, are presented in Table [A5]. Results were very much as expected, and treatment was extremely successful. Only residual levels of COD, which remained about 300 mg/l after biological treatment, were of any concern. It was demonstrated that additions of alkalinity would be necessary if nitrification processes were not to be inhibited, and these were determined accurately by the trials.

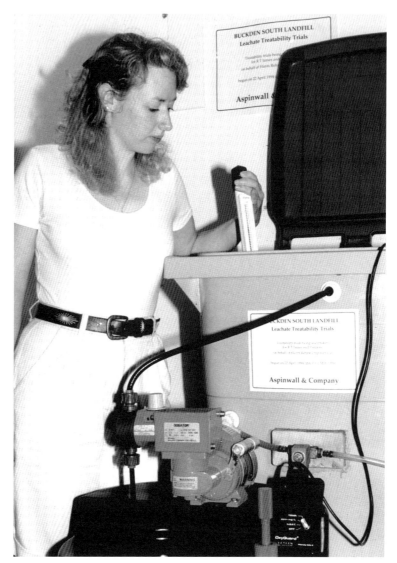

[Figure A16]   Detail of the pilot-scale leachate treatability trial, used to treat leachate at 10 °C at rates of up to 20 litres per day

At the end of the trial (day 64) 100 litres of filtered effluent were supplied to Ozotech Limited, who were to supply the full-scale ozonation unit, for ozonation typical of that which would be provided. This was treated and then sent to Euro Laboratories Limited, of Bedfordshire, in order to determine acute toxicity to rainbow trout, over a 96 hour period, as specified in the proposed effluent consent. The trout were exposed to the following dilutions of effluent (as percent volume per volume, of effluent in water) – 0, 6, 12, 25, 50 and 100 percent. The results indicated that the effluent was not toxic to rainbow trout – no mortalities, or visible sub-lethal effects, were seen in any of the dilutions, up to and including 100 percent effluent. This gave great confidence that the proposed full-scale plant would provide effluent of a quality that would pose no problems with respect to this aspect of the consent.

Table [A5]   Final composition of influent leachate, and settled and filtered
             effluents at equilibrium at the end of the pilot-scale treatability
             trial (day 64)

Determinand:	Final leachate (feed)	Final effluent (unfiltered)	Final effluent (filtered)	Percent removal (unfiltered)
pH-value	7.3	8.3	8.3	–
COD	**841**	**307**	**367**	**63.5**
BOD$_5$	**102**	**15**	**–**	**85.3**
ammoniacal-N	**302**	**0.3**	**<0.3**	**99.9**
chloride	**1560**	**1500**	**1530**	**3.8**
BOD$_{20}$	782	61	<2	92.2
TOC	198	99.4	–	49.8
fatty acids (as C)	<10	<10	–	–
alkalinity (as CaCO$_3$)	2520	690	–	–
conductivity (μS/cm)	8100	7420	–	8.4
nitrate-N	0.3	312	326	–
nitrite-N	<0.1	0.1	<0.1	0
sulphate (as SO$_4$)	15	43	–	0
phosphate (as P)	0.8	1.9	–	0
sodium	1070	1340	1300	–
magnesium	69	81	–	0
potassium	227	228	–	0
calcium	156	163	–	0
chromium	<0.02	<0.02	–	0
manganese	<0.04	0.17	–	0
iron	**2.9**	**1.8**	**–**	**37.9**
nickel	0.012	0.08	–	33.3
copper	0.03	0.04	–	0
zinc	0.08	0.13	–	0
cadmium	<0.01	<0.01	–	0
lead	<0.04	<0.04	–	0
arsenic	0.050	0.017	–	66.0
mercury (μg/l)	<0.1	–	–	–
mecoprop (μg/l)	**3.3**	**0.61**	**–**	**81.5**
isoproturon (μg/l)	**310**	**190**	**–**	**38.7**

Notes:

(a) Results in mg/l except pH-value and conductivity or where shown; – = no result or limit.

(b) **Results presented in bold** for determinands identified as of concern in Table [A4] above.

The full-scale plant treatment process was designed by Aspinwall during late
1994, and was ready for biological commissioning by the end of the year.
(See [Figures A17, A18 and A19]).

The plant comprises twin sequencing batch reactors, each designed for heat
efficiency, and capable of treating a total of up to 200 m^3/d of leachate,
followed by effluent polishing using 2000 m^2 of engineered and planted
reed bed. Each SBR unit is in fact based on an innovative design, with a
'doughnut-shaped' outer aerobic treatment tank, which operates like a racetrack,

and a central settlement tank, in which sludge settles and effluent is clarified for discharge to the main reed bed. An ozonating plant that is capable of dosing up to 150 mg/l of ozone breaks down residual pesticides into smaller organic

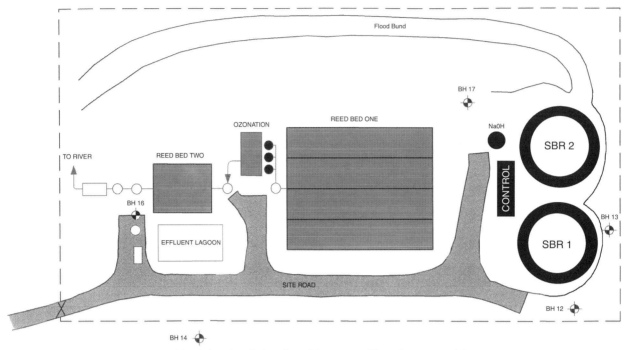

[Figure A17]   Schematic plan of Buckden South Leachate Treatment Plant (not to scale)

[Figure A18]   Buckden Landfill Site, 1994, showing the River Ouse (in flood) to the top left, Buckden South Landfill being restored to the top, Buckden North Landfill being developed in the foreground, and the treatment plant on the left of the picture

molecules, and these are then degraded fully in a final 500 m² reed bed, before effluent is ultimately discharged into the River Ouse. The plant is shown schematically in [Figure A17] and is without doubt the most advanced leachate treatment system yet constructed in the UK. It has operated successfully for nearly eight years.

[Figure A19]   Buckden Leachate Treatment Plant, autumn 2001, showing the two roofed SBR units on the rhs, the main reed beds, the ozonation plant and final reed bed polishing system to the left. The landfill gas power generation system is to the far left

Plant performance

Figures [A20, A21 and A22] present data for treatment performance of the plant during the first six years of operation following commissioning, in terms of removal of COD, BOD_5 and ammoniacal-N.

The plant has performed consistently and reliably since completion of full biological commissioning at the start of February 1995, and all effluent discharged has always met effluent consent values. During a brief period in December 1995, a problem with the alkali dosing plant led to a temporary increase in levels of ammoniacal-N in final effluent (see [Figure A22]), but daily monitoring of the plant was able to identify this and avoid discharge of this water to the Ouse, until the problem was rectified.

[Figures A23 to A26] show more detail of the treatment system, and [Figure A27] shows the final discharge point into the River Ouse.

Table [A6] presents typical operating results as leachate is treated by passage through the full-scale system, and gives typical final effluent composition. Complete nitrification is routinely achieved, and results from monthly use of toxicity testing using the rainbow trout was so successful (no fish ever harmed) that this test was replaced by the much cheaper Microtox® test for routine use, which uses bioluminescent algae to detect toxic effects in effluent. None have yet been observed.

The only determinands which it was evident would be difficult to reduce to initial indicative consented levels were chloride, and COD value. In the light of the successful toxicity testing, the value for chloride was increased to a consented value of 2000 mg/l, which has proved entirely adequate.

Extended operation of the plant has demonstrated that while BOD_5 values in final effluent rarely exceeded 10 mg/l (limit 20 mg/l), COD values were routinely in the range 250 mg/l–350 mg/l, and it would be very difficult to achieve the indicative limit value of 200 mg/l, ultimately proposed to be reduced to 100 mg/l. After detailed discussion with the Environment Agency, therefore, extensive effluent characterisation studies were undertaken on leachate as it passed through the various treatment processes using GC/MS analyses. Initially, these demonstrated that increasing the ozone dose rate did not achieve significant further reduction in COD value – although slightly raised BOD_5 values immediately following ozonation (to be degraded in the final reed bed) did indicate that some COD was being broken down into a more

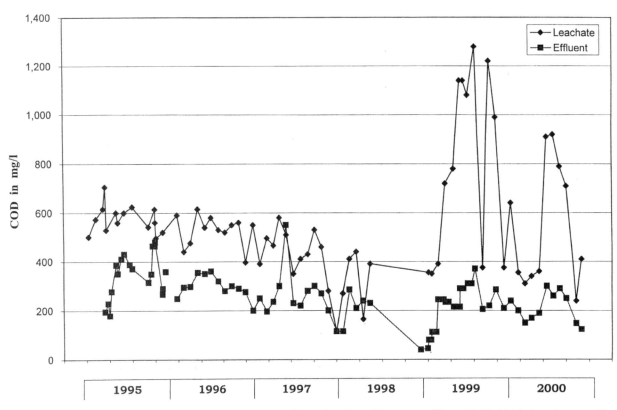

[Figure A20]   Removal of COD achieved by Buckden South Leachate Treatment Plant, 1995–2000 (results in mg/l)

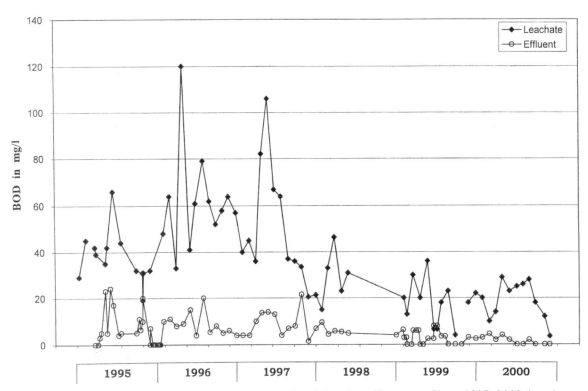

[Figure A21]   Removal of BOD achieved by Buckden South Leachate Treatment Plant, 1995–2000 (results in mg/l)

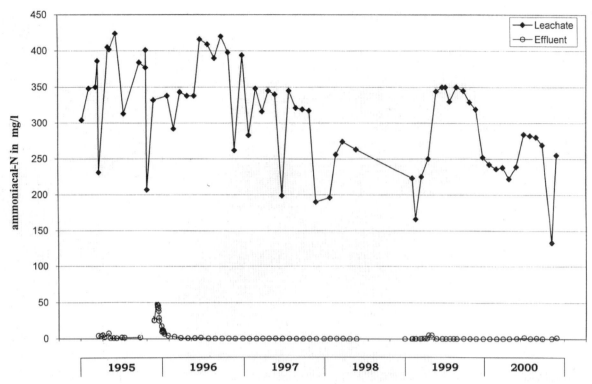

[Figure A22]   Removal of ammoniacal-N achieved by Buckden South Leachate Treatment Plant, 1995–2000 (results in mg/l)

[Figure A23]   Detail of one of the buried Sequencing Batch Reactor (SBR) treatment tanks

[Figure A24]   The main reed bed, with the ozonation plant in the background

[Figure A25]   The ozonation system

[Figure A26]   The ozonation plant and final reed bed

[Figure A27]   The final effluent discharge into the River Ouse, 400 m
from the site

degradable form (see Table [A7]). Since all pesticides were completely
degraded by the lower doses of ozone, this could not be justified, and would
not be successful in meeting the existing COD consent.

Nevertheless, detailed GC/MS studies, using various extraction procedures, only
identified one organic compound of any known toxicity – namely bromal
(tribromacetaldehyde) – and demonstrated that this was not present in either
raw leachate or effluent from the first stage aerobic biological treatment process,
but was a by-product of ozonation.

Results further demonstrated that up to 98 percent of this bromal is degraded by
passage through the final reed bed polishing system. This rapid degradation was
taking place in spite of the facts that reeds have barely begun to colonise the
final gravel bed at this stage – therefore, as reed growth extends, it was
reasonable to assume that [bromal] degradation can only improve.

Additionally, toxicity testing using the rainbow trout (96 hr LC_{50}), and also
using the Microtox$^{®}$ test (15 minutes) did not detect any toxic effects
whatsoever on either organism, when these were exposed to undiluted final
effluent from the treatment process.

On this basis, the Environment Agency were able to increase the consent value
for COD in final effluent to 500 mg/l, which is likely to be adequate for all
future purposes, and has not been exceeded during 8 years of operation of the
plant. The plant continues to treat leachate successfully, controlling leachate
levels around the boundary of the closed, and now graded, capped, and restored
landfill site.

Table [A6]   Typical treatment results for Buckden South Leachate Treatment Plant (May 1995)

Sample:	Leachate	SBR1	SBR2	RB1	+ Ozone	Final
Determinand:						
SS	96	26	14	9	54	4
pH-value	7.2	8.4	7.9	7.9	7.9	7.5
COD	600	424	399	390	395	350
BOD_5	35	4	6	4	11	5
ammoniacal-N	405	1.6	<0.1	0.70	0.8	1.0
chloride	1830	1700	1720	1750	1740	1740
alkalinity (as $CaCO_3$)	2560	1900	1260	1266	637	1270
EC (μS/cm)	–	–	–	9050	–	6510
nitrate-N	0.5	396	409	394	388	382
nitrite-N	<0.05	0.1	0.07	0.5	<0.1	0.4
sulphate (as SO_4)	~50	–	–	–	–	115
phosphate (as P)	0.4	–	–	11.8	–	–
sodium	1100	–	–	–	–	–
magnesium	70	–	–	–	–	–
potassium	190	200	208	195	220	194
calcium	141	134	151	117	128	171
chromium	<0.02	–	–	–	–	–
manganese	0.04	–	–	–	–	–
iron	17.4	<0.06	0.9	0.7	0.6	<0.6
nickel	0.13	–	–	–	–	–
copper	0.03	–	–	–	–	–
zinc	<0.05	0.16	0.21	0.13	0.15	0.14
cadmium	<0.01	–	–	–	–	–
lead	<0.04	–	–	–	–	–
mercury (μg/l)	<0.01	–	–	–	–	–
pesticides (μg/l)						
mecoprop (μg/l)	~200	–	–	–	<0.05	<0.05
isoproturon (μg/l)	~300	–	–	–	3.0	1.9

Notes:

- Results in mg/l except where shown;

- – = no result.

- SBR1 = effluent, SBR tank 1, etc.

- RB1 = effluent, reed bed 1, etc.

Operational costs

Table [A8] is an extremely detailed summary of the performance of the leachate treatment plant at Buckden, and also of electricity costs of operation. Electricity for aeration and for ozonation comprises the main running costs, use of sodium hydroxide for automatic pH-value control, and of phosphoric acid for provision of phosphorus as a nutrient, is relatively small.

Table [A7]   Results from ozonation of effluent from pilot-scale treatment trials
carried out on Buckden South Leachate during early 1995
(sampling date 6 March 1995)

Determinand:	Leachate	Effluent	Ozone 50	Ozone 75	Ozone 100
pH-value	7.1	8.1	7.3	7.2	7.2
COD	615	354	326	311	311
BOD_{20}	119	17	20	20	21
BOD_5	70	3	13	16	14
ammoniacal-N	350	<0.3	0.4	<0.3	0.5
chloride	1490	1680	1670	1660	1660

Notes:

(a) Results in mg/l except pH-value.

(b) Ozone 50 represents 50 mg of ozone dosed per litre, and other samples relate to increased
doses as shown.

In just over seven years for which operational data are presented, approaching
one third of a million cubic metres of leachate has been treated and discharged,
at typical rates of about 150 m^3/d, and at normal unit running costs of between
50 and 70 pence per m^3. During this period, the plant has received many
hundreds of visitors, from the UK and overseas, and continues to be a facility
of which WRG are proud.

Table [A8]   Detailed summary of performance, and of electricity consumption and costs, for the Buckden South Leachate Treatment Plant, by quarter years, January 1995 to March 2002 (all analytical results in mg/l)

Year	Quarter	m^3	COD		BOD$_5$		NH$_4$-N		Electricity Use		
			In	Out	In	Out	In	Out	kWh	£	£/m^3
1995	1st	5150	585	195	39	<1	324	4.1	n.d.	n.d.	n.d.
	2nd	8723	586	323	48	11.0	411	2.8	n.d.	n.d.	n.d.
	3rd	13,248	583	358	38	4.6	348	1.6	n.d.	n.d.	n.d.
	4th	10,492	548	382	28	3.4	327	7.1	n.d.	n.d.	n.d.
1996	1st	7088	502	281	48	2.6	324	8.6	n.d.	n.d.	n.d.
	2nd	6902	578	355	74	9.3	364	1.1	153,217	7587	1.09
	3rd	8295	533	300	64	11.1	406	0.43	162,195	7549	0.91
	4th	8149	478	283	61	5.5	330	0.22	82,605	4229	0.52
1997	1st	9790	479	215	47	4.0	342	0.22	89,940	4520	0.69
	2nd	8250	477	329	73	12.8	300	0.35	95,375	4683	0.57
	3rd	9200	420	250	51	5.5	333	0.17	106,166	5147	0.56
	4th	9200	423	257	30	10.3	275	0.76	102,367	4784	0.52
1998	1st	9000	265	172	23	7.0	226	0.12	96,186	4534	0.50
	2nd	7600	332	227	33	5.3	232	0.15	61,152	3119	0.41
	3rd	0	–	–	–	–	–	–	9003	935	–
	4th	0	–	–	–	–	–	–	2204	2036	–
1999	1st	11,860	365	102	21	7.2	205	0.20	147,395	6857	0.58
	2nd	18,200	880	240	21	4.2	315	1.7	168,041	7693	0.42
	3rd	16,100	843	320	19	3.7	343	0.25	165,424	7572	0.47
	4th	11,040	862	237	7	1.0	331	0.27	158,273	7451	0.67
2000	1st	9300	457	217	17	3.3	243	0.20	155,811	7355	0.79
	2nd	9100	337	170	22	2.7	238	0.33	152,803	6409	0.70
	3rd	9655	873	283	27	0.7	282	0.96	153,601	7142	0.74
	4th	11,710	453	174	11	<1	219	0.56	161,273	6083	0.52
2001	1st	13,490	295	151	9	<2	195	0.16	n.d.	n.d.	n.d.
	2nd	13,660	336	102	9	<2	220	<0.10	n.d.	n.d.	n.d.
	3rd	15,470	580	190	20	<2	216	0.30	n.d.	n.d.	n.d.
	4th	15,030	610		6.5		289		n.d.	n.d.	n.d.
2002	1st	15,810									
12 months to 1 April 2001		43,955	554	209	20	1.3	246	0.61	633,368	25,877	0.59
Consent limits (mg/l)			–	500	–	20.0	–	10.0	–	–	–

n.d. = no data available.

Conclusions

During the last two decades, leachate treatment systems installed at UK landfills have advanced a great deal in sophistication and reliability. Many major landfill sites now include well-designed and engineered plants, that enable leachates to be treated consistently to appropriate and site-specific standards. Many treatment systems use aerobic biological treatment processes, sometimes followed by reed bed polishing, to readily and routinely achieve surface water effluent consent values.

In some circumstances, further effluent polishing using treatment processes such as ozonation, may be justified – for example, to assist in degradation of resistant organic compounds such as some pesticides, as at Buckden. Over seven years of experience by WRG at the site demonstrate that, when needed, these processes can readily be incorporated into leachate treatment schemes. Nevertheless, use of such technology will remain the exception, rather than the rule. At many other sites (about half of the 50 plus landfills where Enviros Aspinwall has designed on-site leachate treatment plants), surface water discharge standards can routinely be achieved simply by use of well-designed SBR systems, often complemented by appropriate reed bed polishing systems.

References

Robinson, H D (2002) Study will predict the impact of changing landfill practice on leachate quality. *Local Authority Waste and Environment*, March 2002, page 17.

Knox, K; Robinson, H D; van Santen, A and Tempany, P R (2000) The occurrence of trace organic compounds in landfill leachates and their removal during on-site treatment. *IWM Scientific and Technical Review*, November 2000, 5–10.

Robinson H D (1996) A review of the Composition of Leachates from Domestic Wastes in Landfill Sites. Report No. CWM 072/95, published by the Waste Technical Division of the Environment Agency, in the series 'The Technical Aspects of Controlled Waste Management', 500pp, August 1996.

Robinson, H D (1989) The development of methanogenic conditions within landfilled wastes, and effects on leachate quality. Paper presented to 'Sardinia 89', the Second International Landfill Symposium *'Landfill Concepts, Environmental Aspects, Lining Technology. Leachate Management, Industrial Waste and Combustion Residues Disposal'*, Porto Conte (Alghero), Sardinia, Italy, 9–13 October 1989, paper XXIX, 9pp.

FILE ARTICLE 5: NEW PRECAUTIONS IN LANDFILL SAFETY

Technology perspective, *The Chemical Engineer*, January 1988.

In the disposal of solid waste from its enormous Ludwigshafen complex, the West German company BASF is attempting to set new standards in landfill safety. It has designed a controllable and repairable landfill system which anticipates probable developments in German law.

The BASF system has been developed with the construction company Bilfinger and Berger at Flotzgrün island, some 40 km upstream of Ludwigshafen on the Rhein. It locates landfill waste above two sheets of impermeable ethylene/bitumen copolymer (ECB) within layers of sand separated by gravel

(Figure [A28]). Water penetrating the landfill reaches the gravel layer above the first sheet and is led away through drainage pipes. The pipes feed to cylindrical sumps [Figure A29], cleverly designed in telescopic units which retain their integrity as the landfill settles. Drainage water is collected and then shipped back to Ludwigshafen where it joins a total flow of some 600,000 m³ per day of water through BASF's waste water treatment plant.

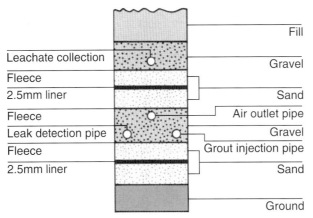

Figure [A28]   The design of the double protective layer

[Figure A29]   Landfill progresses at Flotzgrün

Should the upper plastic liner be damaged, leaks can be detected by pipes set in the gravel 'sandwich' between the two plastic liners. As the site is divided into self-contained units, any unit showing a leak can be identified and individually 'repaired' by injecting grout or bentonite into the space between liners through another pipework system.

The 80 ha Flotzgrün site has been operated by BASF since the late 1960s and currently holds about 9 m tons of landfill. It is estimated that it will serve until the year 2020.

Much of the existing landfill has been dumped according to a simpler regime. Non-degradable material of all kinds was placed on a 50–80 cm thick layer of lime; degradable waste like filter cake from the Ludgwigshafen water treatement plant was placed above a single layer of sheeting from which drainage water was collected. The new system (Figure [A30]) is considerably more expensive to operate but is, BASF managers feel, an inevitable development. Apart from the construction costs, the new system yields some 3000 m^3 of drainage water a month against 200 m^3 from the old system.

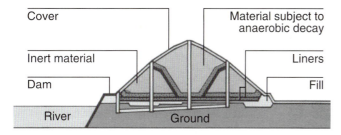

Figure [A30]   Overall cross-section of the Flotzgrun site

BASF scientists were alerted to possible problems with the old system when samples taken from 28 wells dug around the landfill site to a depth of 140 m showed increases in chloride and sulphate levels.

The whole site should eventually accommodate some 21 m^3 of waste and is being capped with topsoil and planted with shrubbery as filling progresses. BASF anticipates making the site available for recreational use in due course.

FILE ARTICLE 6: AN OVERVIEW OF MULTI-CRITERIA ANALYSIS TECHNIQUES

'An overview of multi-criteria analysis techniques', adapted from chapters 4 and 6 of *Multi-criteria Analysis Manual*, Office of the Deputy Prime Minister (2003) [online] **http://www.odpm.gov.uk/stellent/groups/odpm_about/ documents/page/odpm_about_608524.hcsp** [accessed November 2004].

Chapter 4

4.1 Introduction

Although all the techniques to be described in this manual would be widely acknowledged as methods of multi-criteria analysis [MCA], they cover a wide range of quite distinct approaches (in contrast notably to [Cost Benefit Analysis] CBA which is a more unified body of techniques). Some kinds of MCA do not at present offer much help for practical decision taking, but some can be of considerable value. This manual describes and explains these practical techniques, and indicates the types of application in which they may be used.

All MCA approaches make the options and their contribution to the different criteria explicit, and all require the exercise of judgement. They differ however in how they combine the data. Formal MCA techniques usually provide an explicit relative weighting system for the different criteria. The main role of the techniques is to deal with the difficulties that human decision-makers have been shown to have in handling large amounts of complex information in a consistent way.

MCA techniques can be used to identify a single most preferred option, to rank options, to short-list a limited number of options for subsequent detailed appraisal, or simply to distinguish acceptable from unacceptable possibilities.

As is clear from a growing literature, there are many MCA techniques and their number is still rising. There are several reasons why this is so:

- there are many different **types of decision** which fit the broad circumstances of MCA;
- the **time** available to undertake the analysis may vary;
- the amount or nature of **data** available to support the analysis may vary;
- the **analytical skills** of those supporting the decision may vary; and
- the **administrative culture and requirements** of organisations vary.

4.2 Criteria for selecting MCA techniques

Criteria used in this manual for the selection of techniques are:

- internal consistency and logical soundness;
- transparency;
- ease of use;
- data requirements not inconsistent with the importance of the issue being considered;
- realistic time and manpower resource requirements for the analysis process;
- ability to provide an audit trail; and
- software availability, where needed.

4.3 Key features of MCA

Multi-criteria analysis establishes preferences between options by reference to an explicit set of objectives that the decision making body has identified, and for which it has established measurable criteria to assess the extent to which the objectives have been achieved. In simple circumstances, the process of identifying objectives and criteria may alone provide enough information for decision-makers. However, where a level of detail broadly akin to CBA is required, MCA offers a number of ways of aggregating the data on individual criteria to provide indicators of the overall performance of options.

A key feature of MCA is its emphasis on the judgement of the decision making team, in establishing objectives and criteria, estimating relative importance weights and, to some extent, in judging the contribution of each option to each performance criterion. The subjectivity that pervades this can be a matter of concern. Its foundation, in principle, is the decision makers' own choices of objectives, criteria, weights and assessments of achieving the objectives, although 'objective' data such as observed prices can also be included. MCA, however, can bring a degree of structure, analysis and openness to classes of decision that lie beyond the practical reach of CBA.

One limitation of MCA is that it cannot show that an action adds more to welfare than it detracts. Unlike CBA, there is no explicit rationale or necessity for a Pareto Improvement rule that benefits should exceed costs. Thus in MCA, as is also the case with cost effectiveness analysis, the 'best' option can be inconsistent with improving welfare, so doing nothing could in principle be preferable.

4.3.1 Advantages of MCA over informal judgement

MCA has many advantages over informal judgement unsupported by analysis:

- it is open and explicit;
- the choice of objectives and criteria that any decision making group may make are open to analysis and to change if they are felt to be inappropriate;
- scores and weights, when used, are also explicit and are developed according to established techniques. They can also be cross-referenced to other sources of information on relative values, and amended if necessary;
- performance measurement can be sub-contracted to experts, so need not necessarily be left in the hands of the decision making body itself;
- it can provide an important means of communication, within the decision making body and sometimes, later, between that body and the wider community; and
- [when] scores and weights are used, it provides an audit trail.

4.3.2 The performance matrix

A standard feature of multi-criteria analysis is a *performance matrix*, or consequence table, in which each row describes an option and each column describes the performance of the options against each criterion. The individual performance assessments are often numerical, but may also be expressed as 'bullet point' scores, or colour coding. Table [A9] shows a simple example, which we will consider in our detailed introduction to MCDA in Chapter 6. The Table based on an analysis in *Which?* magazine,[18] shows the performance of a number of different toasters in regard to a set of criteria thought to be relevant in a household's choice between different models. These criteria are price, presence of reheat setting, warming rack, adjustable slot width, evenness of toasting, and number of drawbacks. As can be seen, some of these criteria are measured in cardinal numbers (price, number of drawbacks), some in binary terms (a tick indicates presence of a particular feature), and one in qualitative terms (evenness of toasting).

In a basic form of MCA this performance matrix may be the final product of the analysis. The decision makers are then left with the task of assessing the extent to which their objectives are met by the entries in the matrix. Such intuitive processing of the data can be speedy and effective, but it may also lead to the use of unjustified assumptions, causing incorrect ranking of options. ...

In analytically more sophisticated MCA techniques the information in the basic matrix is usually converted into consistent numerical values. In Chapter 6 we show how this can be done using the toaster example.

Table [A9]   Performance matrix

Options	Price	Reheat setting	Warming rack	Adjustable slot width	Evenness of toasting	Number of drawbacks
Boots 2-slice	£18				☆	3
Kenwood TT350	£27	✓	✓	✓	☆	3
Marks & Spencer 2235	£25	✓	✓		★	3
Morphy Richards Coolstyle	£22				☆	2
Philips HD4807	£22	✓			★	2
Kenwood TT825	£30				☆	2
Tefal Thick 'n' Thin 8780	£20	✓		✓	★	5

A tick indicates the presence of a feature. Evenness of toasting is shown in *Which?* on a five-point scale, with a solid star representing the best toaster, and an open star the next best. The family eliminated from consideration all the toasters that scored less than the best or next best.

4.3.3 Scoring and weighting

MCA techniques commonly apply numerical analysis to a performance matrix in two stages:

1   Scoring: the expected consequences of each option are assigned a numerical score on a strength of preference scale for each option for each criterion. More preferred options score higher on the scale, and less preferred options score lower. In practice, scales extending from 0 to 100 are often used, where 0 represents a real or hypothetical least preferred option, and 100 is associated with a real or hypothetical most preferred option. All options considered in the MCA would then fall between 0 and 100.

2   Weighting: numerical weights are assigned to define, for each criterion, the relative valuations of a shift between the top and bottom of the chosen scale.

Mathematical routines, which may be written into computer programmes, then combine these two components to give an overall assessment of each option being appraised. This approach therefore requires individuals to provide those inputs that they are best suited to provide, and leaves computers the task of handling detailed information in a way that is consistent with the preferences that have been revealed by these human inputs.

These approaches are often referred to as *compensatory MCA techniques*, since low scores on one criterion may be compensated by high scores on another. The most common way to combine scores on criteria, and relevant weights between criteria, is to calculate a simple weighted average of scores. The discussion of MCA techniques with explicit weights in this manual concentrates on such simple weighted averages.

Use of such weighted averages depends on the assumption of *mutual independence of preferences*. This means that the judged strength of preference for an option on one criterion will be independent of its judged strength of preference on another. Later in the manual, the assumption will be explained in detail, procedures for testing its validity will be provided, and its role in detecting double-counting of criteria will be explained. Where mutual independence of preferences cannot be established, other MCA procedures are available, although they tend to be more complex to apply.

...

[Reference]

[18] *Which?* (November 1995), published by Consumers' Association, 2 Marylebone Road, London NW1 4DF.

Chapter 6: Multi-criteria decision analysis

6.1 What is MCDA?

A form of MCA that has found many applications in both public and private sector organisations is multi-criteria decision analysis, or MCDA for short (also known as multi-attribute decision analysis, or MADA). This chapter explains what MCDA is and then outlines what is required to carry out such an analysis.

MCDA is both an approach and a set of techniques, with the goal of providing an overall ordering of options, from the most preferred to the least preferred option. The options may differ in the extent to which they achieve several objectives, and no one option will be obviously best in achieving all objectives. In addition, some conflict or trade-off is usually evident amongst the objectives; options that are more beneficial are also usually more costly, for example. Costs and benefits typically conflict, but so can short-term benefits compared to long-term ones, and risks may be greater for the otherwise more beneficial options.

MCDA is a way of looking at complex problems that are characterised by any mixture of monetary and non-monetary objectives, of breaking the problem into more manageable pieces to allow data and judgements to be brought to bear on the pieces, and then of reassembling the pieces to present a coherent overall picture to decision makers. The purpose is to serve as an aid to thinking and decision making, but not to take the decision. As a set of techniques, MCDA provides different ways of disaggregating a complex problem, of measuring the extent to which options achieve objectives, of weighting the objectives, and of reassembling the pieces. Fortunately, various computer programs that are easy to use have been developed to assist the technical aspects of MCDA. ...

6.2 Stages in MCDA

MCDA can be used either retrospectively to evaluate things to which resources have already been allocated, or prospectively to appraise things that are as yet only proposed. Thus, in the following explanations of MCDA, there is no need to distinguish these two uses, though in practice the approach will be realised differently.

[The eight-step process for conducting an MCDA is shown in the box below].

The sections [that] follow describe what has to be done at each step ... Some of the process is technical, but equally important is organising the right people to assist at each stage, and some suggestions about these social aspects will be given in this chapter. A simple example of MCDA will be used to illustrate the stages in selecting a toaster for Fred Jones's family. This modest decision problem would hardly require a full MCDA, but it does provide an illustration unencumbered by the detail and difficulties met in real applications. In the next chapter, real examples will be given. Fred's MCDA appears in the boxes, and the reader wanting a quick introduction to MCDA could just read the boxes. ...

APPLYING MCDA: DETAILED STEPS

1 Establish the decision context.

    1.1 Establish aims of the MCDA, and identify decision makers and other key players.

    1.2 Design the socio-technical system for conducting the MCDA.

    1.3 Consider the context of the appraisal.

2 Identify the options to be appraised.

3 Identify objectives and criteria.

    3.1 Identify criteria for assessing the consequences of each option.

    3.2 Organise the criteria by clustering them under high-level and lower-level objectives in a hierarchy.

4 'Scoring'. Assess the expected performance of each option against the criteria. Then assess the value associated with the consequences of each option for each criterion.

    4.1 Describe the consequences of the options.

    4.2 Score the options on the criteria.

    4.3 Check the consistency of the scores on each criterion.

5 'Weighting'. Assign weights for each of the criteria to reflect their relative importance to the decision.

6   Combine the weights and scores for each option to derive an overall value.

   6.1 Calculate overall weighted scores at each level in the hierarchy.

   6.2 Calculate overall weighted scores.

7   Examine the results.

8   Sensitivity analysis.

   8.1 Conduct a sensitivity analysis: do other preferences or weights affect the overall ordering of the options?

   8.2 Look at the advantage and disadvantages of selected options, and compare pairs of options.

   8.3 Create possible new options that might be better than those originally considered.

   8.4 Repeat the above steps until a 'requisite' model is obtained.

6.2.1 Establish aims of the MCDA, and identify decision makers and other key players

What is the purpose of the MCDA? Get this wrong and you can provide a wonderful analysis for the wrong problem. That's not to say the purpose stays fixed throughout the analysis. As an MCDA progresses new features are often revealed and new issues raised, which may signal a change or shift of aims. Still, the MCDA has to start somewhere, and a statement of initial aims is crucial to formulating the successive stages. After all, MCDA is about determining the extent to which options create value by achieving objectives, and at this stage you face two options: doing the MCDA or not. Choosing to carry out the MCDA means that someone judged the analysis to provide relatively more value than not doing it.

Clarity about the aims of the MCDA helps to define the tasks for subsequent stages and keeps the analysis on track.

AIMS OF THE TOASTER MCDA:

Fred Jones has an old toaster he bought for less than £10 many years ago, and it now toasts unevenly. His youngest daughter Zoe burned her finger when she touched the side of the toaster yesterday as it was toasting his bread at breakfast. Fred can now afford to move up-market and purchase a toaster that will better meet his family's needs. The aim of the MCDA will be to make the best use of the data available to inform the choice of a new toaster.

The first impact for the MCDA of these aims is on the choice of key players to participate in the analysis. A **key player** is anyone who can make a useful and significant contribution to the MCDA. Key players are chosen to represent all the important perspectives on the subject of the analysis. One important perspective is that of the final decision maker and the body to whom that person is accountable, because it is their organisation's values that must find expression in the MCDA. These people are often referred to as **stakeholders**, people who have an investment, financial or otherwise, in the consequences of any decisions taken. They may not physically participate in the MCDA, but their values should be represented by one or more key players who do participate.

STAKEHOLDERS AND KEY PLAYERS:

Members of Fred's family are the stakeholders. His wife Jane wants to consult a neighbour who recently purchased a toaster. She thinks that *Which?* magazine should be consulted too. But she and Fred don't discuss this stage. He intends to ask the advice of his local store whose salesperson he trusts to give impartial recommendations. As we shall see, failure to plan at this stage leads to a problem in the next stage.

No MCDA is ever limited just to the views of stakeholders. Additional key players participate because they hold knowledge and expertise about the subject matter. That includes people within the organisation, and often includes outside experts, or people with no investment in the final decision but who hold information that would assist the analysis. **Designers of the MCDA will need to consider what stakeholders and other key players should be involved, and the extent of their participation in the analysis.**

6.2.2 Design the socio-technical system for conducting the MCDA

When and how are the stakeholders and key players to contribute to the MCDA? That is the social aspect of the design. What form of MCDA is to be used, and how will it be implemented? That is the technical aspect. The two are designed together to ensure they are working in concert to achieve the aims of the MCDA. For example, an MCDA to support a major decision, such as the location of a new airport, will be comprehensive, covering many objectives and criteria, and will involve many interest groups and key players. The complexity of the model will in part dictate who is to contribute, and views expressed by interest groups and key players will influence the complexity of the model. On the other hand, an MCDA to prioritise proposed projects within some governmental unit will involve few if any outsiders and will employ a simpler form of MCDA. There is no one 'best' design. **The social and technical aspects of the system for conducting the MCDA have to be considered together**.

A typical approach to problem solving in the civil service is to hold a series of meetings punctuated in between by staff work, continuing until the task is accomplished. However, an advantage of MCDA is that the process lends itself to designs that are more cost efficient than the typical approach. There are various ways to conduct this.

One approach would be to use *facilitated workshops*. These consist of participants who might be any mix of interest groups and key players. An impartial facilitator guides the group through the relevant stages of the MCDA, carrying out much of the modelling on the spot with the help of computer programs designed for multi-criteria analysis, and with appropriate displays of the model and its results for all to see. Because participants are chosen to represent all the key perspectives on the issues, the workshops are often lively, creative sessions, with much exchange of information between participants whose areas of expertise differ. Indeed, recent research[37] shows that the group can produce judgements that are better than could have been achieved by individuals working separately. Three factors work together to account for this enhanced performance: impartial facilitation, a structured modelling process, and use of information technology to provide on-the-spot modelling and display of results.

An impartial facilitator focuses on process and maintains a task orientation to the work. He or she ensures that all participants are heard, protects minority points of view, attempts to understand what is going on in the group rather than to appraise or refute, attends to relationships between participants, is sensitive to the effects of group processes and intervenes to forward the work of the group[38]. The facilitator of the MCDA assists the groups through the various stages, eliciting relevant expertise and judgements from the participants. By working as a collective, participants often discover interconnections between areas of apparently separate expertise. Each person also sees the larger picture to which the MCDA is addressed, and this larger view can affect individual contributions as their own area is put into perspective. Information technology provides for rapid construction of the MCDA model, facilitates inputting information, and shows results immediately. Because participants see each stage of the model-building process, then are shown how the computer combines the results, the overall model building becomes transparent, contributing to participants' feeling of ownership of the model. Any decisions subsequently made by the responsible person, informed by the MCDA model results, are then understood by those participating, with the result that the decisions are more readily implemented.

DESIGN THE SYSTEM:

Fred neglects this stage, too, and plunges ahead, visiting his local store on the way home from work. His trusted shopkeeper recommends a Philips two-slot, two-slice toaster, which Fred buys. When he arrives home, his wife says that her friend bought a toaster with a warming rack on the top, which could be useful to warm rolls and other small items. That feature is absent on the Philips. Fred realises that his criteria, even toasting and burning risk, are perhaps too restricted, that he should have thought about the process of consulting others about the features of modern toasters. He recognises that his family are all stakeholders, so he decides to explore the features and drawbacks listed in *Which?* and engage the family in discussion before making a final choice. He returns the Philips.

Facilitated workshops might last for only a few hours for relatively straightforward decisions. For complex decisions, two or three-day workshops may be required, or even a series of workshops over a period of several months ... With thoughtful sociotechnical design, workshops can prove to be helpful for a wide range of issues involving resources from tens of thousands of pounds sterling, to hundreds of millions.

6.2.3 Consider the context of the MCDA

What is the current situation? What goals are to be achieved? Could a different frame for the issues and problems provide a recasting of the situation that would make it easier to attain the goals? What strengths can be mobilised to achieve the goals? What weaknesses might impede progress? What opportunities exist now or may appear on the horizon to facilitate progress? What threats could create obstacles? These questions raise concerns that are broader than the aims of the MCDA, but answering them will help to provide a setting for the analysis which will affect many subsequent steps.

Describing the current situation and then being clear about the goals to be achieved establishes the discrepancy between **now** and the **vision for the future** which will clarify the role of the MCDA. Presumably that gap will be closed by those authorised for taking decisions, and allocating resources to help achieve the future state. In what way can the MCDA serve the decision making process? The analysis can be framed in different ways, some more directly supporting the eventual decision, and some less so. The MCDA might be structured to:

- show the decision maker the best way forward;
- identify the areas of greater and lesser opportunity;
- prioritise the options;
- clarify the differences between the options;
- help the key players to understand the situation better;
- indicate the best allocation of resources to achieve the goals;
- facilitate the generation of new and better options;
- improve communication between parts of the organisation that are isolated; or
- any combination of the above.

Looking at strengths, weaknesses, opportunities and threats, a SWOT analysis is particularly useful in developing options. Keeping in mind that options are intended to achieve the goals, participants can be encouraged to generate options that will build on strengths, fix weaknesses, seize opportunities and minimise threats.

CONTEXT:

Fred and his family give some thought to what they really want. The whole family of two adults and three children often eat breakfast together; perhaps they should consider two-slot four-slice toasters in addition to the two-slot two-slice models. On Sunday morning, Fred picks up fresh bagels for Sunday brunch, so an adjustable slot width might be handy. As the family discusses what they might like, Fred suddenly remembers seeing a toaster-oven in the home of an American friend on a recent visit to the United States. Perhaps they should consider that function too, for then the oven could also be used to cook single frozen meals, which would be handy for the occasions when someone comes home late and wants a hot meal.

Other aspects of context concern the larger political, economic, social and technological (PEST) environments in which the analysis is to be conducted. Scenario analysis of how key PEST features might develop in the future, and so affect the ability of the proposed options to achieve the desired future state, sometimes stimulates key players to develop options and consider objectives that would otherwise have been ignored. Scenario analysis[39] can also help participants to acknowledge uncertainty about the future, and thereby make assumptions about outcomes more explicit, thus directing attention at implications which may otherwise be missed.

6.2.4 Identify the options to be appraised

When options are pre-specified, it is tempting to proceed as if that is the final word. Experience shows this is seldom the case. Options are not built in heaven, they are the product of human thought, and so are susceptible to biasing influences. Groups tend to develop fewer options in situations of threat, for example, than when they are facing opportunities. A common error is to attempt to analyse just one option, under the assumption that there is no alternative. But there is always the alternative of continuing as at present, and a proper analysis should be made of that alternative, too.

Options are often formulated on a go/no-go basis. Project funding is often conducted in this way. However, there is an alternative. Bids can be solicited to specify the benefits that would be obtained at different levels of funding. Then, some bids can be rejected altogether, others can be funded at lesser levels, and others at full levels. In this way funding decisions can be made to create more value-for-money. However, this process only works if those bidding for funds have a reasonably clear idea of the main objectives, the value that those allocating the funds wish to create.

IDENTIFY OPTIONS:

Fred's family find that 23 toasters are listed in the November 1995 issue of *Which?*, and they decide to narrow their considerations to the six recommended in the Best Buy Guide. The toaster oven is more than they really need, so the idea is dropped. Fred notes that one toaster not included in the Guide, a long-slot, two-slice toaster, received a 'Best' rating for evenness of toasting, so he wants to include that as a possibility, too, even though *Which?* omitted it because it has too many drawbacks.

In all cases, whether the options are given or have to be developed, those conducting the MCDA should be open to the possibility of modifying or adding to the options as the analysis progresses. A workshop of ICL staff[40] attempted to find a way forward for a project which the managers, who were under severe budget constraints, wanted to cancel, and which the developers wanted to continue because success was only a few months away. The initial MCDA showed what everyone had perceived: that the two options of cancel and continue seemed irreconcilable. After an overnight reflection, participants considered and evaluated new options, each building on the advantages of the previous one, until finally one option was agreed. It involved a joint venture and a partial sell-off, which reduced ICL's cost but actually enhanced their benefits. It was the MCDA that stimulated the creative thought processes of participants, who otherwise would have been resolved to one group losing and the other winning. MCDA can help to create win-win situations.

6.2.5 Identify criteria for assessing the consequences of each option

Assessing options requires thought about the consequences of the options, for strictly speaking it is those consequences that are being assessed, not the options themselves. Consequences differ in many ways, and those ways that matter because they achieve objectives are referred to as criteria, or attributes. Criteria are specific, measurable objectives. They are the 'children' of higher-level 'parent' objectives, who themselves may be the children of even higher-level parent objectives. In choosing a car, you might seek to minimise cost and maximise benefits, two high-level objectives that are in conflict. Benefits might be broken down into categories of safety, performance, appearance, comfort, economy and reliability. Safety could be considered a criterion if you use the rating given by *Which?* of how well the car will protect you in a crash. Alternatively you might wish to disaggregate safety into passive safety and active safety. Passive safety might be treated as a criterion: perhaps a count of the number of features (side bars, roll-over protection, rigid body cage, etc.) would suffice, or it could in turn be further broken down.

A useful distinction is between means and end objectives. Repeatedly ask the question 'Why do you care about that?' and when there is no more to be said, an end objective which is 'fundamental' has been reached. Take the safety criterion for cars. Why do you care about passive safety? It could reduce injuries in the event of a crash. Why do you care about active safety? It increases the chances of avoiding a crash. So for that person reducing injuries and increasing chance of survival are the two fundamental objectives. Or perhaps safety could be considered a fundamental objective if the interpretation of it given in *Which?* magazine is used: how well the car will protect you in a crash.

Which of these many interpretations are susceptible to measurement? The number of passive safety features is easy to count, but it would be a means objective, not an end in itself, so it would not be a fundamental objective. Measurement is often easier for means objectives, yet it is fundamental objectives we may care about. Perhaps experts can provide informed assessments: car safety is rated in *Which? Car* on a single numerical scale. In short, deciding on criteria to incorporate in the MCDA is very much a matter of judgement, and can require some loss in the directness with which the value is expressed in order to facilitate measurement. But as the section below on scoring the options indicates, measurement can include the direct expression of preference judgements, and these may be relatively easy even though no objective measurement is possible.

Criteria express the many ways that options create value. If options are already given, then a 'bottom-up' way to identify criteria is to ask how the options differ from one another in ways that matter. A 'top-down' approach is to ask about the aim, purpose, mission or overall objectives that are to be achieved. Sometimes overall objectives are given. The DETR's new approach to appraisal of transport investments specifies these high-level objectives for transport schemes:

- to protect and enhance the built and natural **environment**;
- to improve **safety** for all travellers;
- to contribute to an efficient **economy**, and to support sustainable economic growth in appropriate locations;
- to promote **accessibility** to everyday facilities for all, especially those without a car; and
- to promote the **integration** of all forms of transport and land use planning, leading to a better, more efficient transport system.

These are further broken down into criteria, some of which are susceptible to numerical measurement, including monetary valuation, others to rating, and some to qualitative description only.

Whose objectives are to be incorporated into the MCDA? Objectives often reflect the core values of an organisation. In a comparison of 18 'visionary' companies with 18 merely 'excellent' companies, Collins and Porras[41] found that:

> Contrary to business school doctrine, 'maximizing shareholder wealth' or 'profit maximization' has not been the dominant driving force or primary objective through the history of the visionary companies. Visionary companies pursue a cluster of objectives, of which making money is only one – and not necessarily the primary one. Yes, they seek profits, but they're equally guided by a core ideology – core values and sense of purpose beyond just making money. Yet, paradoxically, the visionary companies make more money than the more purely profit-driven comparison companies.

Core values for the some of the 18 visionary companies studied by Collins and Porras include being pioneers for an aircraft manufacturer, improving the quality of life through technology and innovation for an electronics manufacturer, technical contribution for a computer manufacturer, friendly service and excellent value for a hotel chain, preserving and improving human life for a medical company, and bringing happiness to millions for an entertainment corporation. These values infuse decision making in the visionary companies, and the authors found many instances in which profits were forgone in order to uphold the values.

Collins and Porras conclude that for these visionary companies, profit is a means to more important ends. In this sense, government departments are no different from the visionary commercial organisations – both exist to create non-financial value; only the source of their funds to do this is different. Thus, **identifying criteria requires considering the underlying reasons for the organisation's existence, and the core values that the organisation serves**.

IDENTIFY CRITERIA:

The Jones family agrees with Fred that evenness of toasting and protection from burned fingers are essential, but four other features are included in the *Which?* review. Nobody has thought of some, like a reheat setting for warming cold toast. And eight potential disadvantages are listed. This is beginning to look unnecessarily complex; after all, it is only a toaster! A brief discussion reduces the criteria to just six: price, reheat setting, warming rack, adjustable slot width, evenness of toasting, and number of disadvantages.

The UK Treasury's 'Green Book' on appraisal and cost benefit analysis states that analysis within government is concerned with effects 'on the national interest.' Of course different institutions might interpret this in different ways, reflecting for example the views of experts, Ministers, senior officials, public opinion, or those directly affected by the decision. For example, the criteria used to capture distributional judgements will vary from case to case, depending upon institutional mindsets and the known preferences of the government of the day. A broadly satisfactory requirement for MCDA is that criteria should be chosen to represent the concerns of people as a whole, and to allow the expression of informed preferences.

6.2.6 Organise the criteria by clustering them under higher-level and lower-level objectives in a hierarchy

Organising the criteria and objectives in this way facilitates scoring the options on the criteria and examining the overall results at the level of the objectives. The most important trade-off between the objectives appears at the top of the hierarchy. This is often between costs and benefits. Thus, the very top objective is the overall result, taking both costs and benefits into account. The next level down would show costs as one objective, and benefits as another. Costs could then be broken down into monetary costs and non-monetary costs, or short-term and long-term, or capital and operating, or any other distinction that captures more conflict between the objectives. The same applies to benefits. Top-level trade-offs are not always between costs and benefits. Other possibilities include risks versus benefits, benefits to consumers versus benefits to suppliers, long-term benefits versus short-term benefits, and so forth. This hierarchical representation is often referred to as a **value tree**.

Figure [A31] shows an illustration of how objectives and criteria for the DETR's new approach to appraisal of transport investments might be represented:

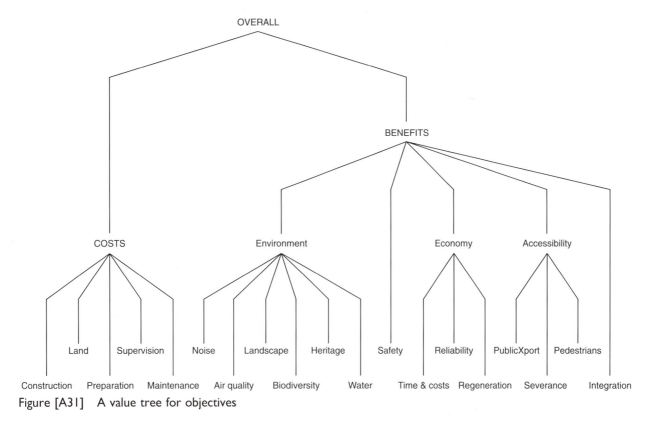

Figure [A31]   A value tree for objectives

The five objectives have been clustered under the higher-level objective 'BENEFITS', the cost of the investment has been separated out of the 'Economy' objective and represented as a separate objective, with its sub-costs represented beneath as criteria. That separation facilitates the display of benefits versus costs for schemes being appraised. There are no sub-objectives for 'Safety' and 'Integration', so those objectives also serve as criteria. This representation is meant as illustration only; it might need modification if MCDA were to be applied.

> **ORGANISE THE CRITERIA:**
>
> The benefits associated with the toasters don't appear to be related to their costs, at least not for the seven toasters on the Jones's short list, so they don't bother with this step. They just want to know what is best overall.

Organising the objectives and criteria in a value tree often highlights the conflict amongst the objectives, and this can lead to refining their definitions. Making the value tree explicit and displaying it may stimulate thinking about new options that could reduce the apparent conflicts between the objectives ... **Iterating back to previous stages is typical in any MCDA.**

6.2.7 Describe the consequences

The easiest approach is to write a simple qualitative description for each option taking into account each criterion. For simpler problems, a performance matrix ... will often suffice. For complex problems that involve a value tree, it may be necessary to construct a separate consequence table for each option, much like the Appraisal Summary Table for the DETR's new approach to appraisal for transport investments. Such a table is structured like the value tree, with separate columns (or in the case of the DETR summary table, rows) for each criterion. The bottom row usually gives the performance measures for that option on the column's criterion. Higher level objectives are shown in rows above the subsidiary criteria, throughout the table.

DESCRIBE THE CONSEQUENCES:

The Jones family copy the data they are interested in directly from *Which?* to give the following performance matrix, which we have already met as an example in Chapter 4.

Options	Price	Reheat setting	Warming rack	Adjustable slot width	Evenness of toasting	Number of drawbacks
Boots 2-slice	£18				☆	3
Kenwood TT350	£27	✓	✓	✓	☆	3
Marks & Spencer 2235	£25	✓	✓		★	3
Morphy Richards Coolstyle	£22				☆	2
Philips HD4807	£22	✓			★	2
Kenwood TT825	£30				☆	2
Tefal Thick 'n' Thin 8780	£20	✓		✓	★	5

A tick indicates the presence of a feature. Evenness of toasting is shown in *Which?* on a five-point scale, with a solid star representing the best toaster, and an open star the next best. The family eliminated from consideration all the toasters that scored less than the best or next best.

6.2.8 Score the options on the criteria

At this point a problem arises. It isn't possible to combine money, ticks, stars and ratings to achieve an overall evaluation of the toasters. However, apples **can** be compared with oranges, and MCDA shows how this is done. The key idea is to construct scales representing preferences for the consequences, to weight the scales for their relative importance, and then to calculate weighted averages across the preference scales. There are many ways to do all this ... [Here], relative preference scales will be illustrated. These are simply scales anchored at their ends by the most and least preferred options on a criterion. The most preferred option is assigned a preference score of 100, and the least preferred a score of 0, much like the Celsius scale of temperature. Scores are assigned to the remaining options so that differences in the numbers represent differences in strength of preference. These are relative judgements comparing differences in consequences, and they are often easier for people to make than absolute judgements. For example, most people would agree that gaining £50,000 would give them pleasure, while losing £50,000 would cause pain, and that the pain is more undesirable than the pleasure is desirable. The pain exceeds the gain. But by how much? Is the pain twice as bad, five times as bad, 10 times as bad, or what? That is a more difficult judgement to make, but most people have some feel for it. And it is this kind of judgement that is required in MCDA. Modelling, such as that done in cost-benefit analysis, can be used for some criteria to assist in the process of converting consequences into scores that are comparable.

What do these preference scores represent? The difference-scaling method results in numbers that represent relative strength of preference. Such a measure expresses the value associated with the option's consequence on a particular criterion. The phrase 'strength of preference' is here used instead of 'value', because the latter is often thought to imply only financial value. However, 'strength of preference' should not be confused with 'preference.' Recall that in decision theory coherent preference logically implies two measurable quantities, probabilities and utilities.[42] Thus, A could be preferred to B if they are equal in value because A is more likely. If strength of preference is to be taken only as a measure of value, then A and B must be assumed to be equally likely. When they aren't, then the uncertainty associated with A and B must be accommodated in some other way, as discussed later, in section 7.4, so that strength of preference measures reflect only relative value.

SCORE THE OPTIONS:

The family uses relative scaling to replace the consequences in their table with scores, with the following result [Table A10]:

[Table A10   Preference scores matrix]

Options	Price	Reheat setting	Warming rack	Adjustable slot width	Evenness of toasting	Drawbacks	Total
Boots 2-slice	100	0	0	0	0	50	35
Kenwood TT350	25	100	100	100	0	80	61
Marks & Spencer 2235	42	100	100	0	100	50	53
Morphy Richards Coolstyle	67	0	0	0	0	100	30
Philips HD4807	67	100	0	0	100	90	49
Kenwood TT825	0	0	0	0	0	90	9
Tefal Thick 'n' Thin 8780	84	100	0	100	100	0	70

Preference scores for price are, of course, the inverse of the prices; the Boots is least expensive, so scores 100, and the Kenwood TT825 the most pricey, so scores 0. All others are scaled between those limits in proportion to the inverse of their prices. For the next three criteria, if the feature is present it scores 100, otherwise 0. There are only two ratings for evenness of toasting, so the better rating scores 100, and the less good one 0. That seemed unfair at first, but later the family realises that the weighting process takes account of the small difference between the top two ratings for evenness. Finally, the family realises that the number of drawbacks can't easily be converted to preferences because it depends on what the drawbacks are; some are more important to them than others. Instead, they simply look at the nature of the drawbacks, and directly assess preference scores to reflect their overall impression of the seriousness of the drawbacks.

Relative scaling is particularly appropriate for comparing several options presented at the same time. Sometimes, however, options are evaluated serially, so that comparison to a standard is required. It is often helpful to use fixed scales for these cases. The zero point for a fixed scale on a given criterion might be defined as the lowest value that would be acceptable – any option scoring less would be rejected outright whatever its scores on other criteria. The 100 point could be defined as the maximum feasible – this would require imagining and defining a hypothetical option as a top scorer.

6.2.9 Check the consistency of the scores on each criterion

This stage is usually accomplished during the process of assessing scores, but is included here separately to emphasise its importance. The method for checking consistency depends on the type of scale used. For the relative scales used in this chapter, the approach is to compare differences on a given scale. If the scale has been constructed properly, then comparing differences was a part of the scoring process, so the scale should be consistent.

In MCDA consistency of preferences is a virtue, and helps to ensure valid results. The initial assessment of scores often reveals inconsistencies, both within and between criteria. Several iterations may be needed until the key players feel that there is sufficient consistency in their preferences. The modelling process actually helps people to attain that goal; consistency is not required to start.

CHECK CONSISTENCY:

The price is consistent with the Joneses' view that their preferences are linear (inversely) over the £18 to £30 range of prices on their list. For example, the difference in preference between the Kenwood and Marks & Spencer toasters is 42 points, the same as between the Marks & Spencer and the Tefal; both pairs differ in price by £5. Giving those two differences in price the same difference in preference means that the Jones family is indifferent between spending £5 more than £20, and £5 more than £25. If the latter increase was considered more onerous than the first, then the first decrement in preference would have been smaller than the second.

6.2.10 Assess weights for each of the criteria to reflect its relative importance to the decision

The preference scales still can't be combined because a unit of preference on one does not necessarily equal a unit of preference on another. Both Fahrenheit and Celsius scales include 0 to 100 portions, but the latter covers a greater range of temperature because a Celsius degree represents nine-fifths more temperature change than a Fahrenheit degree. Equating the units of preference is formally equivalent to judging the relative importance of the scales, so with the right weighting procedure, the process is meaningful to those making the judgements.

ASSIGN WEIGHTS TO THE CRITERIA:

The Joneses decide to allocate 100 points against the criteria as weights. They compare the differences between the most and least preferred toasters for all criteria, and agree that the £12 price difference matters most to them, so give it 30 points. Fred argues that he cares most about evenness of toasting, but Caroline, his elder daughter, points out that the short list is made up of toasters all of which gave above average evenness, so the difference between the top two ratings isn't very important. The family decides the small difference in evenness is about half as important as price, so it gets 15 points. The presence of a warming rack is about equal, so it receives 15 points, too. The adjustable slot seems a little less important than the price difference, but not much, so it is assigned 25 points. With just 15 points left, 10 is given to drawbacks, and 5 to the reheat feature. The ratios of those weights seems about right, though the family feels a little unsure, and Fred is still concerned about the low weight on evenness.

Most proponents of MCDA now use the method of 'swing weighting' to elicit weights for the criteria. This is based, once again, on comparisons of differences: how does the swing from 0 to 100 on one preference scale compare to the 0 to 100 swing on another scale? To make these comparisons, assessors are encouraged to take into account both the difference between the least and most preferred options, and how much they care about that difference. For example, in purchasing a car, you might consider its cost to be important in some absolute sense. However, in making the choice of a particular car, you might already have narrowed your choice to a shortlist of, say, five cars. If they only differ in price by £200, you might not care very much about price. That criterion would receive a low weight because the difference between the highest and lowest price cars is so small. If the price difference was £2000, you might give the criterion more weight unless you are very rich, in which case you might not care.

There is a crucial difference between measured performance and the value of that performance in a specific context. Improvements in performance may be real but not necessarily useful or much valued: an increment of additional performance may not contribute a corresponding increment in added value.

Thus, the weight on a criterion reflects both the range of difference of the options, and how much that difference matters. So it may well happen that a criterion which is widely seen as 'very important' – say safety – will have a similar or lower weight than another relatively lower priority criterion – say maintenance costs. This would happen if all the options had much the same level of safety but varied widely in maintenance costs. Any numbers can be used for the weights so long as their ratios consistently represent the **ratios** of the valuation of the **differences** in preferences between the top and bottom scores (whether 100 and 0 or other numbers) of the scales which are being weighted.

Implementing the swing weighting method with a group of key players can be accomplished by using a 'nominal-group technique.' First, the one criterion with the biggest swing in preference from 0 to 100 is identified. If the MCDA model includes only a few criteria, then the biggest swing can usually be found quickly with agreement from participants. With many criteria, it may be necessary to use a paired-comparison process: compare criteria two at a time for their preference swings, always retaining the one with the bigger swing to be compared to a new criterion. The one criterion emerging from this process as showing the largest swing in preference is assigned a weight of 100; it becomes the standard to which all the others are compared in a four-step process. First, any other criterion is chosen and all participants are asked to write down, without discussion, a weight that reflects their judgement of its swing in preference compared to the standard. If the criterion is judged to represent half the swing in value as the standard, for example, then it should be assigned a weight of 50. Second, participants reveal their judged weights to the group (by a show of hands, for example, against ranges of weights: 100, 90s, 80s, 70s, etc.) and the results are recorded on a flip chart as a frequency distribution. Third, participants who gave extreme weights, high and low, are asked to explain their reasons, and a general group discussion follows. Fourth, having heard the discussion, a subset of participants makes the final determination of the weight for the criterion.

Who makes up the subset? Usually the decision maker, or those representing the decision maker, or those participants (often the most senior ones) whose perspectives on the issues enable them to take a broad view, which means that they can appreciate the potential tradeoffs among the criteria. Thus, the final determination of the weights is informed by a group discussion that started from knowledge of where everybody stood, uninfluenced by others. The process also engages those closest to the accountable decision maker in making judgements that are uniquely that person's responsibility, whether or not they are expressed numerically.

The setting of weights brings to the fore the question of whose preferences count most. ... [This] manual can go no further than identify this as an issue which should be recognised explicitly rather than implicitly. The choice is ultimately political and may depend on the context. However it noted that a broadly satisfactory criterion which appears to underlie many CBA valuations is that they should reflect the informed preferences of people as a whole, to the extent that these preferences and the relative importance of the criteria can be expressed in numbers. This might often be a sound aspiration for MCDA. However it may not be an aspiration which is shared, at least initially, by all those who might expect to be consulted about a particular application.

The process of deriving weights is thus fundamental to the effectiveness of an MCDA. Often they will be derived from the views of a group of people. They might reflect a face-to-face meeting of key stakeholders or people able to articulate those stakeholders' views, in which weights are derived individually, then compared, with an opportunity for reflection and change, followed by broad consensus. If there is not a consensus, then it might be best to take two or more sets of weights forward in parallel, for agreement on choice of options can sometimes be agreed even without agreement on weights. Even if this does not lead easily to agreement, explicit awareness of the different weight sets and their consequences can facilitate the further search for acceptable compromise.

In MCDA the meaning of weights, despite these difficulties, is reasonably clear and unambiguous. The concept of a 'weight' takes on different meanings with other MCA methods used. It always needs to be handled with care.

6.2.11 Calculate overall weighted scores at each level in the hierarchy

This is a task for computers, though a calculator is sometimes sufficient. The overall preference score for each option is simply the weighted average of its scores on all the criteria. Letting the preference score for option i on criterion j be represented by s_{ij} and the weight for each criterion by w_j, then for n criteria the overall score for each option, S_i, is given by:

$$S_i = w_1 s_{i1} + w_2 s_{i2} + ... + w_n s_{in} = \sum_{j=1}^{n} w_i s_{ij}$$

In words, multiply an option's score on a criterion by the importance weight of the criterion, do that for all the criteria, then sum the products to give the overall preference score for that option. Then repeat the process for the remaining options.

6.2.12 Calculate overall weighted scores

The theory of MCDA makes clear that the simple weighted averaging calculation shown above is justified only if one particular condition is met: all the criteria must be **mutually preference independent.** This is a straightforward idea, simpler and less restrictive than real-world independence or statistical independence. It means that the preference scores assigned to all options on one criterion are unaffected by the preference scores on the other criteria. Some examples might be instructive. Two criteria can be causally linked in the real world, creating statistical correlation between the scores on the two criteria, yet be preference independent. Cars with well-appointed interiors are generally more expensive; price and poshness are positively correlated. However, most people generally prefer nicer interiors and less pricey cars. Preference scores can be given for cars' interiors without knowing what the cars cost, and for price without knowing how well appointed the interiors are. Preferences are mutually independent even though correlation exists in the real world. Take another example: choice of a main dish for an evening meal at a restaurant. In forming preferences for the main dishes on the menu, most people don't first look at the wine list.

Preferences for main dishes are independent of the wines. But preferences for wines may well depend on the main dish. Thus, preference for main dishes is independent of preference for wines, but preference for wines is not independent of preference for main dishes. Yet what goes on in the kitchen is quite independent of the wine cellar. So, a one-way non-independence of preferences exists, even though there is independence in the real world.

Incidentally, the family did not notice that the Philips toaster is in every respect at least as good as the Kenwood TT825, and on three criteria better. [Thus] the Philips toaster is said to 'dominate' the Kenwood TT825, so the latter could have been eliminated from the analysis at this point.

Failure of mutual preference independence, if it hasn't been caught when the criteria are being formed, usually is discovered when scoring the options. If the assessor says that he or she can't judge the preference scores on one criterion without knowing the scores on another criterion, then preference dependence has been detected. This often happens because of double counting ... if two criteria really mean the same thing, but have been described in a way that apparently is different, then when the scores are elicited the assessor will often refer back to the first criterion when assessing the second. That is a signal to find a way to combine the two criteria into just one that covers both meanings.

CALCULATE OVERALL WEIGHTED SCORES:

Caroline is just learning to use a spreadsheet, so she sets up the following table [Table A11] and inputs the formula for calculating the total: a simple weighted average, where the scores in each row are multiplied by the column weights expressed as decimals (e.g. 30 as 0.30) and the products summed to give the total weighted score for each toaster.

Table A11   Calculating overall scores

Options	Price	Reheat setting	Warming rack	Adjustable slot width	Evenness of toasting	Drawbacks	Total
Boots 2-slice	100	0	0	0	0	50	**35**
Kenwood TT350	25	100	100	100	0	80	**61**
Marks & Spencer 2235	42	100	100	0	100	50	**53**
Morphy Richards Coolstyle	67	0	0	0	0	100	**30**
Philips HD4807	67	100	0	0	100	90	**49**
Kenwood TT825	0	0	0	0	0	90	**9**
Tefal Thick 'n' Thin 8780	84	100	0	100	100	0	**70**
Weights	**30**	**5**	**15**	**25**	**15**	**10**	

Sometimes mutual preference independence fails because one or more options score so poorly on a given criterion that scores on other criteria can't compensate. For example, people who toast breads of varying thickness may feel that the low score for a toaster with fixed slot width can't be compensated for by other features associated with high scores. This may be a signal to reject all fixed-slot toasters outright. That has the advantage of restoring mutual preference independence for the remaining options. But if that can't be done, then MCDA can still accommodate the failure by using slightly more complex mathematics, usually including multiplicative elements along with the simple weighted averaging model of this section. Multiplying preference scores causes an overall low preference if either of the two numbers multiplied together is low; this aspect of the model is non-compensatory. However, for most applications in government, particularly when fixed scales are used with the lowest position defined as the minimum acceptable, value above the minimum is additive, so the simple compensatory model is adequate.

6.2.13 Examine the results: Agree the way forward or make recommendations

The top-level ordering of options is given by the weighted average of all the preference scores. These total scores also give an indication of how much better one option is over another. Thus, if the total scores for options A, B and C are 20, 60 and 80, the difference in overall strength of preference between A and B is twice as large as that between B and C. Another, slightly awkward, way to express this is that compared to B, A is twice as less preferred as C is more preferred. Another useful display of overall results is to move down a level in the value tree and display the options in a two-dimensional plot to show the main trade-offs. If costs and benefits constitute the next level down, then a graph of benefits versus costs can be instructive, for it essentially shows a relative value-for-money picture. The outer surface of the plot gives the most cost-effective options.

Options appearing on the outer surface are said to 'dominate' options inside because they are both more beneficial and less costly.

An MCDA can yield surprising results that need to be digested before decisions are taken. It may be necessary to establish a temporary decision system to deal with unexpected results and to consider the implications of new perspectives revealed by the MCDA. This temporary system consists of a series of working meetings which eventually produce recommendations to the final decision making body. At the working meetings, participants are given the task of examining the MCDA results, testing the findings for their validity, working though the possible impacts for the organisation, and formulating proposals for the way forward. When MCDA throws up surprises, it is tempting to ignore this post-MCDA stage, to demean the analysis, and find some other basis for supporting decisions. But it is important to recognise that if discrepancies between MCDA results and people's intuitions have not been explored, the MCDA model was not 'requisite'.[43] Exploring the discrepancies does not mean the sense of unease will go away; on the contrary, it could be heightened if the MCDA is found to be sound, but the message it is conveying is unpleasant or unwelcome. A period of working through the results ensures that subsequent decisions are taken with full awareness of possible consequences.

EXAMINE RESULTS:

The family are surprised to see the Tefal as the overall winner with a total score of 70. It wasn't among the best buys *Which?* recommended! Tom, the middle child, asks if it is possible just to show the benefits separate from the costs. Caroline sets the weight on costs to zero, recalculates the benefits, then plots the overall benefits versus the costs. This shows that compared to the lower-cost Boots toaster, the Tefal provides much more benefit for little extra cost, whereas the extra benefit from the Kenwood 350 is not as cost efficient. The Tefal dominates the remaining four; it is both more beneficial and less costly. The family find the graph helps to give an overall picture of the toasters [Figure A32].

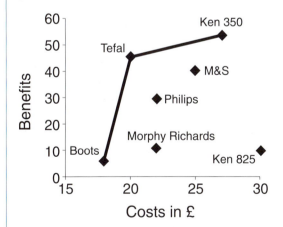

[Figure A32   Graph of benefits versus costs for toasters]

6.2.14 Conduct sensitivity analysis: Do other preferences or weights affect the overall ordering of the options?

Sensitivity analysis provides a means for examining the extent to which vagueness about the inputs or disagreements between people makes any difference to the final overall results. Especially for appraisal of schemes or projects that attract public interest, the choice of weights may be contentious.

Experience shows that MCDA can help decision makers to reach more satisfactory solutions in these situations.

First, interest groups can be consulted to ensure that the MCDA model includes criteria that are of concern to all the stakeholders and key players. Second, interest groups often differ in their views of the relative importance of the criteria, and of some scores, though weights are often the subject of more disagreement than scores. Using the model to examine how the ranking of options might change under different scoring or weighting systems can show that two or three options always come out best, though their order may shift. If the differences between these best options under different weighting systems are small, then accepting a second-best option can be shown to be associated with little loss of overall benefit. The reason this is usually not apparent in the ordinary thrust of debate between interest groups is that they focus on their differences, and ignore the many criteria on which they agree. Third, sensitivity analyses can begin to reveal ways in which options might be improved ... **There is a potentially useful role for sensitivity analysis in helping to resolve disagreements between interest groups.**

CONDUCT SENSITIVITY ANALYSES:

Fred thinks if more weight had been given to evenness of toasting, his original purchase, the Philips, would look better overall because it received a best rating for evenness in *Which?*. Caroline doubles the weight on evenness from 15 to 30. That does indeed improve the overall benefits of the Philips, but as can be seen in the new graph [Figure A33], the Tefal's overall score increases as well. Fred then realises this has to be the case because the Tefal received a best rating for evenness as well, so increasing the weight on that criterion also helps the Tefal. Now the Tefal dominates all the toasters except the Boots! The family decides that their original set of weights better reflects their values.

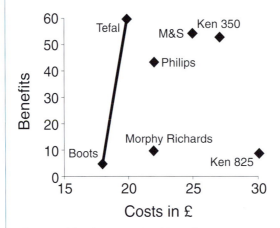

[Figure A33   New graph of benefits versus costs for toasters]

6.2.15 Look at the advantages and disadvantages of selected options, and compare pairs of options

Many analyses can be carried out to deepen understanding of the issues associated with the MCDA. These extra analyses are easily conducted with the help of computer programs designed to implement MCDA; more is said about the programs and the analyses in the chapter on results. In addition to automatically plotting graphs like those above, these programs enable users quickly to establish the advantages and disadvantages of each option, and to compare options. An advantage is a high score on a heavily weighted criterion;

a high score on a relatively unimportant criterion isn't really an advantage because it doesn't contribute to overall preference. A disadvantage is a low score on an important criterion. Disadvantages are important because they reduce the overall preference, whereas low scores on unimportant criteria don't. Understanding the advantages and disadvantages helps to point to areas where options might be capable of improvement.

Comparing options is particularly useful when one option is naturally a standard. Big differences in preference scores between pairs of options on important criteria can be identified quickly, aiding the process of developing new and better options. Another helpful comparison is between the option that scores best on benefits, and the one that is least costly.

6.2.16 Create possible new options that might be better than those originally considered

The key differences between pairs of options might point to ways of generating a new option. For example, comparison of the most beneficial option with the least costly one may show how to create a new option with many, though not quite all, of the benefits of the most beneficial option, but is less costly. Sometimes this is accomplished by reducing the benefits, and thus the cost, on those criteria that do not carry much weight. Reducing the cost in this way may more than compensate for the loss of benefit, giving an option that is quite beneficial without being too costly.

If new options are generated, add them to the list of options and score the new option on all criteria. If relative scaling was used and the new option is least preferred on some criteria or most preferred on others, then it is easier to assign scores less than 0 or more than 100, respectively, so that weights do not have to be changed. An important feature of MCDA is that if the new option provides no information about the existing options and criteria, then nothing already completed has to be changed. It is only necessary to add one more preference score for each criterion, and that's all.

6.2.17 Repeat the above steps until a 'requisite' model is obtained

A requisite model[44] is one that is just good enough to resolve the issues at hand. Less work should be done for modest problems that are of less importance, when time is short and resources are limited. The Joneses' toaster analysis was more than requisite, although they learned some unexpected things from it. Many organisations spend unnecessary amounts of time gathering information, refining inputs and modelling. A key question to ask of any activity forming a part of an analysis is, 'Will this activity, whatever its outcome, make any difference to a decision?' If not, then the activity is not worth pursuing.

An important characteristic of MCDA models is that they are often remarkably insensitive to many scores and weights. This is easily demonstrated in sensitivity analysis, but until this insensitivity has been experienced, people often find it difficult to live with rough-and-ready inputs. 'Come back in six months after we have gathered more data' is a common reaction to the suggestion that an up front MCDA will help to show what data matter, that sensitivity analysis will reveal the tolerance of final results to substantial imprecision in the many of the inputs. Many people have experience of models where precision matters. The reason that imprecision is so well tolerated in MCDA models is that the scores on many of the criteria will show high statistical correlation, and thus the weights on those criteria can be distributed amongst the correlated criteria in any way. In addition, changes in scores on individual criteria are often swamped by the scores for the same options on other criteria. Thus, the structure of any model that includes many criteria creates this lack of sensitivity. As experience is gained of MCDA, models become simpler and increasingly requisite.

THE JONESES' TOASTER PROBLEM:

These last three steps weren't carried out by the family; that would have been overkill for a simple problem, and little if anything more would have been learned. So what happened? Fred and Jane decided to take a look at the Tefal and Kenwood 350 toasters. The length of the Tefal worried Jane when she saw it; *Which?* magazine hadn't given dimensions, only small photographs of the toasters. She hadn't realised how long a long-slot toaster is, and she thought that the Kenwood was rather bulky. The kitchen is small, she hates clutter on the work surfaces, and the footprint of the Kenwood seemed aesthetically more acceptable. Fred was quite taken with the warming rack as his toast often goes cold while he drinks his coffee and reads his newspaper. In the end, they decided to buy the Kenwood. The MCDA had helped them in many ways, but the final decision was theirs to take.

6.3 Uncertainty, risk and MCDA

Because the theoretical roots of MCDA are in decision theory, it is possible to accommodate uncertainty in a coherent way. The formally correct approach is to construct a decision tree, show the consequences at the end of the tree, and then use MCDA to generate a single overall preference score for each consequence. Those scores are then folded back through the decision tree by applying the expected utility rule. This results in a probability-weighted score for each option, providing a clear overall preference ordering of the options. Unfortunately, textbooks on decision analysis treat uncertainty and multiple objectives in separate chapters, leaving the reader with the task of combining the two in real problems. Keeney and Raiffa[45] however, provide several real cases in which the two were integrated, and Hammond, Keeney and Raiffa[46] neatly blend the two approaches in a presentation that is intuitive and appealing (see their chapters 7 and 8).

A closely related approach is useful if there is so much uncertainty about the future, and so many possible events, that the decision tree becomes unmanageable. Under those circumstances, constructing several bounding scenarios about possible developments for key external events is easier than constructing a complex decision tree. A separate MCDA is carried out assuming each scenario. This could be accomplished within one value tree by allowing the scenarios to be represented at the highest-level branches, with exactly the same structure of objectives and criteria under each parent scenario. The scenarios can be weighted to reflect their relative plausibility or likelihood of occurring. Sensitivity analyses can be carried out on the weights given to the scenario branches to see the effects of the scenarios on the overall ordering of the options.

Uncertainty might attend only one or two criteria. If this is the case, then separate modelling of the uncertainties might allow preference scoring to be based on expected values. For example, suppose there is uncertainty about the net number of people who will experience an increase in noise if a new road scheme is built. It is usually possible to establish the minimum, most likely, and maximum number of people affected. Those three values could be used to define a triangular probability distribution that roughly describes the uncertainty. The mean of that distribution, given by the average of the three values, would then be used in the MCDA. If subsequent sensitivity analysis showed that this value could crucially affect a decision, then more sophisticated modelling of the uncertainty would be called for. This would require the careful assessment of a subjective probability distribution; see chapter 8 of Clemen[47] for how this can be accomplished.

Another approach is to include a 'confidence' criterion in the value tree, defined as the probability that the other benefits will be obtained. Assessing a score on this criterion for a given option amounts to judging the probability that the option will create the value assessed on the other criteria. This probability is then converted to a negative penalty score which becomes disproportionately more negative the smaller the probability. Placing more or less weight on this confidence criterion provides a means of seeing how sensitive the results are to more or less concern for risk. ...

Finally, some groups will wish to express the risk they feel is associated with the options, where for them risk is not just a probability or simply a reflection of uncertainty. For example, experts and the public at large may not share the same view of what risk means to them.[48] Again, it might be possible to establish a 'confidence' criterion, but options are then assessed for their relative risk, however the key players wish to define risk, using preference scores rather than probabilities.

Risk and uncertainty can be difficult topics, and there are many ways to take them into account in any MCDA. Professional help may be needed to find the best way of accommodating these concerns.

[References and notes]

37   See, for example, Regan-Cirincione, P. (1994) 'Improving the accuracy of group judgment: A process intervention combining group facilitation, social judgment analysis, and information technology', *Organizational Behavior and Human Decision Processes*, 58, 246–70.

38   Phillips, L. D. and Phillips, M. C. (1993) 'Facilitated work groups: theory and practice', *Journal of the Operational Research Society*, 44(6), 533–49.

39   Schoemaker, P. J. H. (1991) 'When and how to use scenario planning: a heuristic approach with illustration', *Journal of Forecasting*, 10, 549–564.

40   Hall, P. (1986) 'Managing change and gaining corporate commitment', *ICL Technical Journal*, 7, 213–27.

41   Collins, J. C. and Porras, J. I. (1996) *Built to Last: Successful Habits of Visionary Companies*, Century Limited, London, p.8.

42   Strictly speaking, utility is a measure of both subjectively judged value and the assessor's attitude toward risk. Utilities are properly assessed using hypothetical wagers, which invoke the assessor's risk propensities. Because the direct scaling method used here does not involve wagers, the resulting numbers are measures only of value.

43   'Requisite' is defined in section 6.2.17.

44   Phillips, L. D. (1984) 'A theory of requisite decision models', *Acta Psychologica*, 56, 29–48.

45   Keeney, R. L. and Raiffa, H. (1976) *Decisions with Multiple Objectives: Preferences and Value Trade-offs* Wiley, New York, reprinted, Cambridge University Press, 1993.

46   Hammond, J. S., Keeney, R. L., and Raiffa, H. (1999) *Smart Choices: a Practical Guide to Making Better Decisions* Harvard University Press, Boston, MA.

47   Clemen, R. T. (1996) *Making Hard Decisions, an Introduction to Decision Analysis* second edition, Duxbury Press Belmont, CA.

48   A good discussion of risk from the perspective of social scientists can be found in Pidgeon, N., Hood, C., Jones, D., Jones, B. and Gibson, R. (1992). 'Risk perception', *Risk: Analysis, Perception and Management* (pp. 89–134), The Royal Society, London.

APPENDIX 2: THE GROUNDWATER DIRECTIVE LISTS

LIST I

1 Organohalogen compounds and substances that may form such compounds in the aquatic environment
2 Organophosphorus compounds
3 Organotin compounds
4 Substances that possess carcinogenic, mutagenic or teratogenic properties in or via the aquatic environment
5 Mercury and its compounds
6 Cadmium and its compounds
7 Mineral oils and hydrocarbons
8 Cyanides

LIST II

1 The following metals and their compounds

    a Zinc
    b Copper
    c Nickel
    d Chrome
    e Lead
    f Selenium
    g Arsenic
    h Antimony
    i Molybdenum
    j Titanium
    k Tin
    l Barium
    m Beryllium
    n Boron
    o Uranium
    p Vanadium
    q Cobalt
    r Thallium
    s Tellurium
    t Silver

2 Biocides and their derivatives not appearing on List I
3 Substances that have a deleterious effect on the taste and/or odour of groundwater, and compounds liable to cause formation of such substances in such water and to render it unfit for human consumption

4   Toxic or persistent organic compounds of silicon, and substances that may cause the formation of such compounds in water, excluding those that are biologically harmless or are rapidly converted in water into harmless substances

5   Inorganic compounds of phosphorus and elemental phosphorus

6   Fluorides

7   Ammonia and nitrites

Notes:

List I substances that present a low risk on the basis of toxicity, persistence and bio-accumulation should be included in List II.

List II substances that are carcinogenic, mutagenic or teratogenic are to be included in category 4 of List I.

REFERENCES

Anon (1996) 'Illinois study on bioaerosol impact on community looks favourable to composters', *Biocycle*, May.

Ballast Phoenix (2004) [Online] http://www.ballastphoenix.co.uk/home.html [Accessed 7 September 2004].

Bomel (2004) *Mapping Health and Safety Standards in the UK Waste Industry*, Research Report 240, London, HMSO.

Breum, N. O., Nielsen, B. H., Nielsen, E. M., Midtgaard, U. and Poulsen, O. M. (1997) 'Dustiness of compostable waste: a methodological approach to quantify the potential of waste to generate airborne microorganisms and endotoxin', *Waste Management and Research*, vol. 15, no. 2, pp. 169–88.

British Standards Institution (1990) *Specification for Lightweight Aggregates For Masonry Units and Structural Concrete*, ISBN 0 580 19024 2.

British Standards Institution (2002) *BS 3797: Specification for composted materials*, PAS 100, ISBN 0 580 40590 7.

Brown, K. A., Smith, A., Burnley, S. J., Campbell, D. J. V., King, K. and Milton, M. J. T. (1999) *Methane Emissions from UK Landfills*, AEA Technology, AEAT-5217.

Browne, M. L., Ju, C. L., Recer, G. M., Kallenbach, L. R., Melius, J. M. and Horn, E. G. (2001) 'A prospective study of health symptoms and *Aspergillus fumigatus* spore counts near a grass and leaf composting facility', *Compost Science & Utilization*, vol. 9, pp. 242–9.

Bünger, J., Antlauf-Lammers, M., Schulz, T. G., Westphal, G. A., Müller, M. M., Ruhnau, P. and Hallier, E. (2000) Health complaints and immunological markers of exposure to bioaerosols among biowaste collectors and compost workers, *Occupational and Environmental Medicine*, vol. 57, pp. 458–64.

Burnley, S. J., Coleman, T. and Gronow, J. R. (1999) 'The impact of the Landfill Directive on strategic waste management in the UK' in *Landfill Processes and Waste Pre-treatment Vol IV*, Proceedings Sardinia 1999 Seventh International Waste Management and Landfill Symposium, Cagliari, Italy.

Cimino, J. A. (1975) 'Health and Safety in The Solid Waste Industry' *American Journal of Public Health* vol. 65, no. 1, pp. 38–46

CIWM (1996) *The Role and Operation of the Flushing Bioreactor*, Northampton, Chartered Institution of Wastes Management.

CIWM (2003) *Energy from Waste: A Good Practice Guide*, Northampton, Chartered Institution of Wastes Management.

CIWM (undated) *The Biological Techniques in Solid Waste Management and Land Reclamation*, Northampton, Chartered Institution of Wastes Management.

Council of the European Communities (1975) 75/442/EEC [The Waste Framework Directive].

Council of the European Communities (1979) 'Council Directive on protection of groundwater against pollution caused by certain dangerous substances, 80/68/EEC', *Official Journal* L 020, 26/01/1980 P. 0043–0048 [the Groundwater Directive].

Council of the European Communities (1991) 'Directive on hazardous waste 91/689/EEC', *Official Journal* L 377, 31/12/1991 [the Hazardous Waste Directive].

Council of the European Union (1999) 1999/31/EC [the Landfill Directive].

Council of the European Union (2003) 'Council Decision establishing criteria and procedures for the acceptance of waste at landfills pursuant to Article 16 of and Annex II to Directive 1999/31/EC'. *Official Journal* L11, pp. 27–49 [Waste Acceptance Criteria].

Cox, C. S. and Wathes, C. M. (eds) (1995) *The Bioaerosols Handbook*, Boca Raton, Lewis Publishers.

Crook, B., Higgins, S. and Lacey, J. (1987) *Airborne Micro-organisms Associated with Domestic Waste Disposal*, Final Report to the HSE Contract Number: 1/MS/126/643/82.

Dalton, P. (2002) 'Odour, irritation and perception of health risk', International Archives of Occupational and Environmental Health (Berlin), vol. 75, no. 5, pp. 283–90.

Danneberg, G., Grueneklee, E., Seitz, M., Hartung, J. and Driesel, A. J. (1997) 'Microbial and endotoxin immissions in the neighborhood of a composting plant', *Annals of Agricultural and Environmental Medicine*, vol. 4, pp. 169–73.

DEFRA (2004a) *Municipal Waste Management Survey 2002/3*, London, HMSO; [Online] http://www.defra.gov.uk/environment/statistics/wastats/mwb0203/wbtables.htm [Accessed November 2004].

DEFRA (2004b) *Review of Environmental and Health Effects of Waste Management – Municipal Solid Waste and Similar Wastes*, London, HMSO.

DEFRA (2005) *Municipal Waste Management Statistics*, London, The Stationery Office [Online] http://www.defra.gov.uk/environment/statistics/wastats/index.htm [Accessed 24 August 2006].

DEFRA (2006) *Municipal Waste Management in the European Union*, London [Online] http://www.defra.gov.uk/environment/statistics/waste/kf/wrkf08.htm [Accessed 1 December 2006].

Déportes, I., Benoit-Guyod, J.-L. and Zmirou, D. (1995) 'Hazard to man and the environment posed by the use of urban waste compost – a review', *The Science of the Total Environment*, vol. 172, pp. 197–222.

de Smet, B. (ed.) (1990) *Life Cycle Analysis for Packaging: Environmental Assessment*, Proceedings of a specialised workshop, Proctor and Gamble Technical Centre, Leuven, The Netherlands, 24–25 September.

DETR (2000) *Waste Strategy 2000*, London, The Stationery Office [Online] http://www.defra.gov.uk/environment/waste/intro.htm [Accessed December 2004].

DoE (Department of the Environment) (1988) *The Licensing of Waste Facilities*, Waste Management Paper 4, 2nd edition, London, HMSO, ISBN 0117521574.

DoE (Department of the Environment) (1989) *The Control of Landfill Gas*, Waste Management Paper 27, London, HMSO.

DoE (Department of the Environment) (1997) *Landfill Design, Construction and Operational Practice*, Waste Management Paper 26B, London, HMSO.

Douwes, J., Wouters, I. and Dubbeld, H. (2000) 'Upper airway inflammation assessed by nasal lavage in compost workers: a relation with bioaerosol exposure', *American Journal of Industrial Medicine* (New York, NY), vol. 37, pp. 459–68.

Elliott, P., Briggs, D., Morris, S., de Hoogh, C., Hurt, C., Jensen, T. K., Maitland, I., Richardson, S., Wakefield, J. and Jarup, L. (2001) 'Risk of adverse birth outcomes in populations living near landfill sites', *British Medical Journal*, vol. 323, no. 7309, pp. 363–8.

Environment Agency (1994) *National Household Waste Analysis Project: Phase 2: Chemical Analysis Data Volume 3,* CWM 087/94, Bristol, The Environment Agency.

Environment Agency (2002) *Interpretation of the Engineering Requirements of Annex I of the Landfill Directive*, Regulatory Guidance Note 6.0 v3.0, Bristol, The Environment Agency.

Environment Agency (2003a) *Monitoring of Particulate Matter in Ambient Air Around Waste Facilities*, Technical Guidance Document M17, Environment Agency R&D Project P1-441, Bristol, The Environment Agency.

Environment Agency (2003b) *Hazardous Waste: Interpretation of the Definition and Classification of Hazardous Waste*, Bristol, The Environment Agency, ISBN 1 84432 130 4.

Environment Agency (2003c) *Hydrogeological Risk Assessments for Landfills and the Derivation of Groundwater Control and Trigger Levels,* LFTGN01, Bristol, The Environment Agency.

Environment Agency (2006) *Strategic Waste Management Information 2002–03*, Bristol, The Environment Agency [Online] http://www.environment-agency.gov.uk/subjects/waste/1031954/315439/923299/?version=1&lang=_e [Accessed 28 August 2006].

Environment Agency/SEPA (2002) *Guidance on Landfill Gas Flaring*, Bristol, The Environment Agency, ISBN 1844320278.

Epstein, E. (1994) 'Composting and bioaerosols', *Biocycle*, vol. 35, no. 1, pp. 51–8.

Epstein, E., Wu, N., Youngberg, C. and Croteau, G. (2001) 'Dust and bioaerosols at a biosolids composting facility', *Compost Science and Utilization*, vol. 9, no. 3, pp. 250–5.

Fava, L., Denison, R., Jofies, B., Curran, M. A., Vigon, B., Selke, S. and Barnum, J. (eds) (1991) *A Technical Framework for Life Cycle Assessments*, Report of the workshop organised by the Society of Environmental Toxicology and Chemistry, Smugglers Notch, Vermont, USA.

Ferreira, J. J., Boocock, M. G. and Gray, M. I. (2004) *Review of the Risks Associated with Pushing and Pulling Heavy Loads*, Research Report 228, London, HMSO.

Fredrickson, L. (1992) *Safety in Recycling Facilities: A Resource for Operators*, Minnesota Pollution Control Agency.

Gale, P. (2002) *Risk Assessment: Use of Composting and Biogas Treatment to Dispose of Catering Waste Containing Meal*, Final Report to the Department for Environment, Food and Rural Affairs, Swindon, WRc plc.

Gilbert, E. J., Ward, C. W., Crook, B., Fredrickson, J., Garrett, A. J., Gladding, T. L., Newman, A. P., Olsen, P., Reynolds, S., Stentiford, E., Sturdee, T. and Wheeler, P. (1999) 'The research and development of a suitable monitoring protocol to measure the concentration of bioaerosols at distances from composting sites' leading to the publication *Standardised Protocol for the Sampling and Enumeration of Airborne Microorganisms at Composting Facilities*, ISBN: 0 9532546 2 3, The Composting Association.

Gilbert, J. and Gladding, T. L. (2004) *Health and Safety at Composting Sites: A Guide for Site Managers*, The Composting Association, ISBN: 0 9532546 9 0.

Giroud, J. P., Khatami, A. and Badu-Tweneboath, K. (1989) 'Evaluation of the rate of leakage through composite liners', *Geotextiles and Geomembranes*, vol. 8, pp. 78–111.

Gladding, T. L. (2002) 'Health risks of materials recovery facilities', pp. 53–72 in *Environmental and Health Impact of Solid Waste Management Activities*, Issues in Environmental Science and Technology 18, Royal Society of Chemistry.

Gladding, T. L. (2004) 'An assessment of the risks to human health of materials recovery facilities: a framework for decision makers', Final Report, *Environment Agency Contract No: P1–214.*

Gladding, T. L. and Coggins, P. C. (1997) 'Exposure to microorganisms and health effects of working in UK materials recovery facilities – a preliminary report', *Annals of Agricultural and Environmental Medicine*, vol. 4, pp. 137–41.

Gladding, T. L., Thorn, J. and Stott, D. (2003) 'Organic dust exposure and work-related effects among recycling workers', *American Journal of Industrial Medicine*, vol. 42, no. 6, pp. 584–91.

GMB (Public Services Union) (2003) *Waste Industry Health and Safety: Refuse Collection*, London, GMB National Office [Online] www.gmb.org.uk [Accessed 28 September 2004].

Gregory, R., Gillett, A. and Blowes, J. (2003) 'UK landfill gas generating set emissions' in *Proceedings Sardinia 2003 Ninth International Waste Management and Landfill Symposium*, Cagliari, Italy.

Gustavsson, P. (1989) 'Mortality among workers at a municipal waste incinerator', *American Journal of Industrial Medicine*, vol. 15, pp. 245–53.

Hall, D. H. and Marshall, P. (1991) 'The role of construction quality assurance in the installation of geomembrane liners', *The Planning and Engineering of Landfills*, pp. 187–92, Midland Geotechnical Society.

Hall, D. H., Drury, D., Smith, J., Potter, H. and Gronow, J. (2003) 'Predicting the groundwater impact of modern landfills: Major developments of the approach to landfill risk assessment in the UK (Landsim 2.5)' in *Proceedings Sardinia 2003 Ninth International Waste Management and Landfill Symposium*, Cagliari, Italy.

Heldal, K., Eduard, W. and Bergum, M. (1997) 'Bioaerosol exposure during handling of source separated household waste', *Annals of Agricultural and Environmental Medicine*, vol. 4, pp. 45–51.

Herr, C. E. W., zur Nieden, A., Jankofsky, M., Stilianakis, N. I., Boedeker, R.-H. and Eikmann, T. F. (2003) Effects of bioaerosol polluted outdoor air on airways of residents: a cross sectional study, *Occupational and Environmental Medicine*, vol. 60, pp. 336–42.

Higgins, S., Crook, B. and Lacey, J. (1987) 'Airborne microorganisms in refuse disposal facilities', *Aerosols, Their Generation, Behaviour and Applications*, pp. 141–4.

HMSO (1998) *The Groundwater Regulations 1998*, Statutory Instrument 1998 no. 2746.

HMSO (2003) *The Animal By-Products Regulations 2003*, Statutory Instrument 2003 no. 1482.

Hobson, A., Frederickson, J. and Dise, N. (2004) 'Emissions of CH_4 and N_2O from composting: comparing mechanically turned windrow and vermicomposting systems', *Biodegradable and Residual Waste Management Conference*, Leeds, CalRecovery Europe Ltd.

Hryhorczuk, D., Curtis, L., Scheff, P., Chung, J., Rizzo, M., Lewis, C., Keys, N. and Moomey, M. (2001) 'Bioaerosol emissions from a suburban yard waste composting facility', *Annals of Agricultural and Environmental Medicine*, vol. 8, pp. 177–85.

HSE (1990) *COSHH Assessments: A Step-by-Step Guide to Assessment and the Skills Needed for it*, London, HMSO.

HSE (2005) *EH40: Occupational Exposure Limits* [the most current up to date version can be found via HSE Publications].

Hummel, J. (2004) *Meeting Statutory Recycling Targets Through Cost Effective Kerbside Expansion*, Milton Keynes, The Open University.

IEERR (Interagency Energy and Environmental Research Report) (1995) *Environmental, Economic, and Energy Impacts of Materials Recovery Facilities: A MITE Programme Evaluation*, Report No. EPA/600/R-95/125, NREL/TP430–8130.

ISO (1997) BS EN ISO 14040: 1997, *Environmental Management – Life Cycle Assessment – Principles and Framework*, International Organisation for Standardization.

Ivens, U. I., Breum, N. O., Ebbehoj, N., Nielsen, B. H., Poulsen, O. M. and Wurtz, H. (1997) 'Exposure-response relationship between gastrointestinal problems among waste collectors and bioaerosol exposure', *Scandinavian Journal of Work and Environmental Health*, vol. 25, no. 3, pp. 238–45.

Ivens, U. I., Ebbehoj, N., Nielsen, B. H., Poulsen, O. M. and Wurtz, H. (1999) 'Gastrointestinal symptoms among waste recycling workers', *Annals of Agricultural and Environmental Medicine*, vol. 4, pp. 153–7.

IWM (Institute of Wastes Management) (2000) *Materials Recovery Facilities*, IWM Business Services Ltd, ISBN 0 902944 57 6.

Jarup, L., Briggs, D., de Hoogh, C., Morris, S., Hurt, C., Lewin, A., Maitland, I., Richardson, S., Wakefield, J. and Elliot, P. (2002) 'Cancer risks in populations living near landfill sites in Great Britain', *British Journal of Cancer*, vol. 86, pp. 1732–6.

Kerrell, E. (1991) *Hazardous Household Waste: The SORT Survey Results*, Leeds, SWAP Ltd.

Knox, K. (1985) 'Leachate production, control and treatment' in *Hazardous Waste Management Handbook*, ed. A. Porteous, London, Butterworths.

Krause, T. (2000) 'OSHA sizes up ergonomics: how will it fit the recycling industry?', *Resource Recycling*, January.

Lacey, J. and Dutkiewicz, J. (1994) 'Bioaerosols and occupational lung disease', *Journal of Aerosol Science*, vol. 25, no. 8, pp. 1371–404.

Lavoie, J. and Alie, R. (1997) 'Determining the Characteristics to be Considered from a Worker Health and Safety Standpoint in Household Waste Sorting and Composting Plants', *Annals of Agricultural and Environmental Medicine*, vol. 4, pp. 123–8.

Lavoie, J. and Guertin, S. (2001) 'Evaluation of health and safety risks in municipal solid waste recycling plants', *Journal of the Air and Waste Management Association* (Pittsburgh, PA), vol. 51, pp. 352–60.

Lewin, K., Blakey, N. C., Turrell, J. A., van der Sloot, H., Collins, R., Bradshaw, K., Russell, A. and Harrison, J. (1995) *The Properties and Utilisation of MSW Incineration Residues,* ETSU report B/RR/00368/REP.

Lewin, K., Blakey, N. C., Turrell, J. A., van der Sloot, H., Collins, R., Bradshaw, K., Russell, A. and Harrison, J. (1996) *The Properties and Utilisation of MSW Incineration Residues: Phase 2*, ETSU report B/RR/00368/REP/2.

London Borough of Barnet (2004) [online] http://www.barnet.gov.uk/recycling [Accessed 28 September 2004].

Malmros, P. (1988) 'The working conditions at Danish sorting plants', ISWA Proceedings, 1, pp. 487–94.

Malmros, P. (1990) 'Problems with the working environment in solid waste treatment' *The National Labour Inspection of Denmark*, Report no. 10.

Malmros, P. (1992) *Get Wise on Waste – A Book about Health and Waste-Handling*, Danish Working Environment Service.

Malmros, P., Pedersen, L. F., Trøster, K. and Tønning, K. (1993) '3-delt indsamlingssystem for dagrenovation Rapport fra et forsøg i Fåborg og Herning Kommuner', *Miljøprojekt* no. 214.

McDougall, F., White, P., Franke, M. and Hindle, P. (2001) *Integrated Solid Waste Management: A Life Cycle Inventory,* 2nd edition, Oxford, Blackwell Science.

Mozzon, D., Brown, D. A. and Smith, J. W. (1987) 'Occupational exposure to airborne dust, respirable quartz and metals arising from refuse handling, burning and landfilling', *American Industrial Hygiene Association Journal*, vol. 48, no. 2, pp. 111–16.

Neef, A., Albrecht, A., Tilkes, F. and Harpel, S. (1999) 'Measuring the spread of airborne microorganisms in the area of composting sites', *Schriftenr Ver Wasser Boden Lufthyg*, vol. 104, pp. 655–4.

Nielsen, E. M., Breum, N. O., Nielsen, B. H., Wurtz, H., Poulsen, O. M. and Midtgaard, U. (1997) 'Bioaerosol exposure in waste collection: a comparative study on the significance of collection equipment, type of waste, and seasonal variation', *Annals of Occupational Hygiene*, vol. 41, pp. 325–44.

Nielsen, E. M., Nielsen, B. H. and Breum, N. O. (1995) 'Occupational bioaerosol exposure during collection of household waste', *Annals of Agricultural and Environmental Medicine*, vol. 2, pp. 53–9.

Nilsson, P. (1996) 'Waste collection technologies – with special focus on occupational health' in First International Course on Bioaerosol Exposure and Health Problems in Relation to Waste Collection and Recycling, 6–10 May, Lyngby, Denmark.

Poll, A. J. (2004) *The Composition of Municipal Solid Waste in Wales*, Government of the National Assembly of Wales [online] http://www.wales.gov.uk/subienvironment/toc-e.htm#r [Accessed 7 September 2004].

Poulsen, O. M., Breum, N. O., Ebbehøj, N., Hansen, A. M., Ivens, U. I., Lelieveld, D. V., Malmros, P., Matthiasen, L., Nielsen, B. H., Nielsen, E. M., Schibye, B., Skov, T., Stenbaek, E. I. and Wilkins, C. K. (1995a) 'Sorting and recycling of domestic waste. Review of occupational health problems and their possible causes', *Science of the Total Environment*, vol. 168, no. 1, pp. 33–56.

Poulsen, O. M., Breum, N. O., Ebbehøj, N., Hansen, A. M., Ivens, U. I., Lelieveld, D. V., Malmros, P., Matthiasen, L., Nielsen, B. H., Nielsen, E. M., Schibye, B., Skov, T., Stenbaek, E. I. and Wilkins, C. K. (1995b) 'Collection of domestic waste: review of occupational health Problems and their possible causes', *The Science of the Total Environment*, vol. 170, pp. 1–19.

Rabl, A., Spadaro, J. V. and McGavran, P. D. (1998) 'Health risks of air pollution from incinerators: a perspective', *Waste Management and Research*, vol. 16, pp. 365–88.

Rao, C. Y., Burge, H. A. and Chang, J. C. (1996) 'Review of quantitative standards and guidelines for fungi in indoor air', *Journal of the Air and Waste Management Association*, vol. 46, no. 9, pp. 899–908.

Rapiti, E., Sperati, A., Fano, V., Dell'Orco, V. and Forastiere, F. (1997) 'Mortality among workers at municipal waste incinerators in Rome: a retrospective cohort study', *American Journal of Industrial Medicine*, vol. 31, pp. 659–61.

Robinson, H. (1995) 'A review of landfill leachate composition in the UK', *IWM Proceedings,* January 1995, Northampton, Institute of Waste Management.

Robinson, H. and Latham, B. (1993) 'Timescale for completion', *The Surveyor*, 6 May.

Roth, T. (1991) *Sickerwasser aus der Bioabfallkompostierung – Moglichkeiten de Behandlung und Entsorgung in einem dezentralen Anlagensystem,* Dissertation, GH Kassel, Witzenhausen.

Rylander, R. (1997a) 'Endotoxins in the environment: a criteria document', *International Journal of Occupational Medicine and Environmental Health*, vol. 3, pp. S1-S48.

Rylander, R. (1997b) 'Investigation of the relationship between disease and airborne (1(r)3)- b-D-glucan in buildings', *Mediators of Inflammation*, vol. 6, pp. 275–7.

Rylander, R. (1999) 'Indoor air-related effects and airborne $(1\rightarrow3)$-β-D-glucan', *Environmental Health Perspectives*, vol. 107(S3), pp. 501–3.

Rylander, R., Persson, K., Goto, H., Yuasa, K. and Tanaka, S. (1992) 'Airborne B(1(r)3)-glucan may be related to symptoms in sick buildings', *Indoor Environment*, vol. 1, pp. 263–7.

Scheepers, P. T. J. and Bos, R. P. (1992) 'Combustion of diesel fuel from a toxicological perspective: II Toxicity', *International Archives of Occupational and Environmental Health*, vol. 64, pp. 163–77.

Schilling, B., Heller, D., Graulich, Y. and Gottlich, E. (1999) 'Determining the emission of microorganisms from biofilters and emission concentrations at the site of composting areas', *Schriftenr Ver Wasser Boden Lufthyg*, vol. 104 pp. 685–701.

SEPA (2002) *Framework for Risk Assessment for Landfill Sites: The Geological Barrier, Mineral Layer and the Leachate Sealing and Drainage System,* Stirling, Scottish Environment Protection Agency.

Sigsgaard, T. (1990) 'Respiratory impairment among workers in a garbage-handling plant', *American Journal of Industrial Medicine*, vol. 17, pp. 92–3.

Sigsgaard, T., Abel, A. and Donbaek, L. (1994b) 'Lung function changes among recycling workers exposed to organic dusts', *American Journal of Industrial Medicine*, vol. 25, pp. 69–72.

Sigsgaard, T., Bach, B., Taudorf, E., Malmros, P. and Gravesen, S. (1990) 'Accumulation of respiratory disease among employees in a recently established plant for sorting refuse', *Ugeskr Laeger*, vol. 152, pp. 2485–8.

Sigsgaard, T., Hansen, J. C., Malmros, P. and Christiansen, J. V. (1996) 'Work related symptoms and metal concentration in danish resource recovery workers' in *Biological Waste Treatment*, London, James & James.

Sigsgaard, T., Malmros, P., Nersting, L. and Petersen, C. (1994a) 'Respiratory disorders and atopy in Danish refuse workers', *American Journal of Respiratory and Critical Care Medicine*, vol. 149, pp. 1407–12.

Slater, R., Davies, P. and Gilbert, E. (2005) *The State of Composting in the UK 2003/2004*, Wellingborough, The Composting Association, ISBN 0 954 7797 1 1.

Swan, J. R. M., Kelsey, A., Crook, B. and Gilbert, E. J. (2003) 'Occupational and environmental exposure to bioaerosols from composts and potential health effects: a critical review of published data', *HSE Research Report*, vol. 130, ISBN 0 7176 2707 1.

Swan, J., Unwin, J., Stagg, S., Plant, N. and Crook, B. (2004) *Exposure of Workers to Toxic Gases and Bioaerosols on Landfill Sites*, Sheffield, Health and Safety Laboratory MIC/2004/03.

TBRA (1999) *Waste Sorting Plants: Guidance for Protection Measures*, Bundesarbeitsblatt, Germany, Ministry of Work, TBRA 210.

US Environmental Protection Agency (1993) *Public Health, Occupational Safety and Environmental Concerns in Municipal Solid Waste Recycling Operations*, Report no. EPA/600/R-93/122.

Ushikoshi, T. et al. (2001) 'Operating results and dioxins removal technology for leachate treatment by reverse osmosis system', *Proceedings Sardinia 2001 Eighth International Waste Management and Landfill Symposium*, Cagliari, Italy.

Van der Sloot, H. A., Kosson, D. S., Eighmy, T. T., Comans R. N. J. and Hjelmar, O. (1994) 'An approach towards international standardisation: a concise scheme for testing granular waste leakability' in *Waste Material in Construction and Waste*, Amsterdam, Elsevier Science.

Vasel, J.-L., Rodriguez-Ruiz, L. and Juspin, H. (2003) 'A database for characterizing landfills and leachates', *Proceedings Sardinia 2003 Ninth International Waste Management and Landfill Symposium*, Cagliari, Italy.

Wheeler, P. A., Stewart, I., Dumitrean, P. and Donovan, B. (2001) 'Health effects of composting: a study of three compost sites and review of past data', *Environment Agency R&D Technical Report*, P1–315/TR.

Wilkins, C. K. (1997) 'Gaseous organic emissions from various types of household waste', *Annals of Agricultural and Environmental Medicine*, vol. 4, pp. 87–9.

Wouters, I. M., Douwes, J., Doekes, G., Thorne, P. S., Brunekreef, B. and Heederik, D. J. J. (2000) 'Increased levels of markers of microbial exposure in homes with indoor storage of organic household waste', *Applied and Environmental Microbiology*, vol. 66, pp. 627–31.

Wouters, I. M., Spaan, S., Doekes, G. and Heederik, D. (2003) 'Bioaerosol exposure and respiratory and systemic health effects in organic domestic waste and green waste composting' in *Man and His Waste: Bioaerosol Exposure and Respiratory Health Effects in Waste Management*, Thesis Utrecht University, ISBN: 90 393 3513 3.

Wurtz, H. (1996) 'Biological and biochemical analysis of bioaerosols' *First International Course on Bioaerosol Exposure and Health Problems in Relation to Waste Collection and Exposure*, Denmark.

ACKNOWLEDGEMENTS

Grateful acknowledgement is made to the following sources for permission to reproduce material within this product.

Tables

Table 2: National Household Waste Analysis Project (1994) The Environment Agency; *Table 11:* Poulsen, O.M. et al (1995) 'Sorting and recycling of domestic waste', *The Science of the Total Environment*, Vol. 168. With permission from Elsevier Science; *Table 12: Monitoring of particulate matter in ambient air around waste facilities*, Technical Guidance Document M17 (2003) The Environment Agency; *Table 13: Specifications for Composted Materials* PAS100 (2002). British Standards Institution; *Table 14:* Hobson, A.M., Frederickson, J. and Dise, N.B. (2004) Emissions of CH_4 and N_2O from composting: Comparing mechanically turned windrow and vermicomposting systems. In proceedings: *Treatment of Biodegradable and Residual Waste*, Harrogate; *Table 15:* Roth, T. (1991) Sickerwasser aus der Bioabfallkompostierung; *Table 16:* Waste Incineration (England and Wales) Regulations 2002. Crown copyright material is reproduced under Class Licence Number C01W0000065 with the permission of the Controller of HMSO and the Queen's Printer for Scotland; *Table 20: Review of Environmental and Health Effects on Waste Management - Municipal Solid Waste and Similar Wastes* (2004) Department for the Environment, Food and Rural Affairs. Crown copyright material is reproduced under Class Licence Number C01W0000065 with the permission of the Controller of HMSO and the Queen's Printer for Scotland; *Table 21: Review of Environmental and Health Effects on Waste Management - Municipal Solid Waste and Similar Wastes* (2004) Department for the Environment, Food and Rural Affairs. Crown copyright material is reproduced under Class Licence Number C01W0000065 with the permission of the Controller of HMSO and the Queen's Printer for Scotland; *Table 27:* Guidance on Landfill Gas flaring (2002) Scottish Environment Protection Agency; *Table 28:* Knox, K. (1985) Leachate production, control and treatment, in (ed) Porteous, A., Hazardous Waste Management Handbook, Butterworths; *Tables 32, 33 and 34:* Interpretation of the Engineering Requirements of Annex I of the Landfill Directive (2002) Environment Agency; *Table 35:* Landfill Design, Construction and Operational Practice, Waste Management Paper 26B (1997) Department of the Environment. Crown copyright material is reproduced under Class Licence Number C01W0000065 with the permission of the Controller of HMSO and the Queen's Printer for Scotland.

Figures

Figure 4: Epstein, E. et al (2001) Effect of dust control at a composting site on the release of Aspergillus fumigatus, *Compost Science and Utilization*, Vol. 9(3), JG Press, Inc; *Figure 6:* 'Landfill Design, Construction and Operational Practice', Waste Management Paper 26B (1997) Department of Environment. Crown copyright material is reproduced under Class Licence Number C01W0000065 with the permission of the Controller of HMSO and the Queen's Printer for Scotland ;*Figure 7:* Waste Management Paper No 27 'The Control of Landfill Gas: A Technical Memorandum on the Control of Landfill Gas'. Crown copyright material is reproduced under Class Licence Number C01W0000065 with the permission of the Controller of HMSO and the Queen's Printer for Scotland; *Figure 8:* Knox, K. (1985) 'Leachate production, control and treatment', in *Hazardous Waste Management Handbook*, (ed) Porteous, A., Butterworths; *Figs 10 & 11:* Hall, D.H. et al (2003) 'Predicting the groundwater impact of modern landfills: Major developments of the approach to landfill risk assessment in the UK' (Landsim 2.5), *Proceedings Sardinia 2003, Ninth International Waste Management and Landfill Symposium*; *Figure 14:* Fave, J. et al (1991) *A Technical Framework for Life Cycle Assesments*, Report of the workshop organised by the Society of Environmental Toxicology and Chemistry.

Appendix

Article 1: Maris, A. (2000) The next big thing, *Resource*, April 2000. By permission of Resource Publishing Ltd

Article 2: Pearce, F. (1997) Burn Me, *New Scientist*, 22 November 1997

Article 3: Robinson, H. et al. (2003) Remediation of leachate problems at Arpley Landfill site, Warrington, Cheshire, *Scientific & Technical Review*, December 2003. Copyright © Howard Robinson

Article 4: Robinson, H., Walsh, T. and Carville, M. (2002) Advanced leachate treatment at Buckden Landfill, Huntingdon, Chartered Institute of Wastes Management (CIWM) Annual Conference, 2002. By permission of Howard Robinson

Article 5: Technology in Perspective (1988) The Chemical Engineer, January 1988, Institute of Chemical Engineers

Article 6: DTLR Multi-Criteria Analysis Manual, ODPM, © Crown copyright material is reproduced with the permission of the Controller of HMSO and Queen's Printer for Scotland.

Illustrations

Figure A1: Copyright © Bath & N E Somerset Council

Figures A3 & A5: Copyright © Getty Images.

INDEX

T308 COURSE TEAM

Stephen Burnley   *course team chair/author Wastes/Modelling*
Ernie Taylor   *course manager*
Rod Barratt   *author, Air*
Sylvan Bentley   *picture researcher*
Sophia Braybrooke   *editor*
Philippa Broadbent   *buyer, materials procurement*
Rozy Carleton   *course secretary*
David Cooke   *author, Wastes/Modelling*
Daphne Cross   *assistant buyer, materials procurement*
Jonathan Davies   *graphic designer*
Tony Duggan   *learning projects manager*
Jim Frederickson   *critical reader, Wastes*
Toni Gladding   *author, Wastes*
Keith Horton   *author, Project Guide; critical reader, Noise*
Karen Lemmon   *compositor*
James McLannahan   *critical reader, general, Water*
Lara Mynors   *media project manager*
Suresh Nesaratnam   *author, Water*
John Newbury   *critical reader*
Stewart Nixon   *software designer*
Janice Robertson   *editor*
David Sharp   *critical reader, Noise*
Lynn Short   *software designer*
Shahram Taherzadeh   *author, Noise*
Howie Twiner   *graphic artist*
James Warren   *author/critical reader, Air*

In addition the course team wishes to thank the following for reviewing the material:

External assessor   Professor E. Stentiford, University of Leeds
Block 1   Brian Buckley, consultant

MAIN TEXTS OF T308

An Introduction to Modelling
Block 1   Potable Water Treatment
Block 2   Managing Air Quality
Block 3   Assessing Noise in Our Environment
Block 4   Solid Wastes Management